First published 1983 by CRC Press
Taylor & Francis Group
6000 Broken Sound Parkway NW, Suite 300
Boca Raton, FL 33487-2742

Reissued 2018 by CRC Press

Library of Congress Cataloging in Publication Data
Main entry under title:

Bioceramics of calcium phosphate.

Bibliography.
Includes index.
1. Dentistry—Ceramics. 2. Calcium
phosphate. 3. Teeth. I. Groot, Klaas de.
RK655.B56 617.6'95 82-4346
ISBN 0-8493-6456-6 AACR2

A Library of Congress record exists under LC control number: 82004346

Publisher's Note
The publisher has gone to great lengths to ensure the quality of this reprint but points out that some imperfections in the original copies may be apparent.

Disclaimer
The publisher has made every effort to trace copyright holders and welcomes correspondence from those they have been unable to contact.

ISBN 13: 978-1-315-89103-3 (hbk)
ISBN 13: 978-1-351-07013-3 (ebk)

Visit the Taylor & Francis Web site at http://www.taylorandfrancis.com and the
CRC Press Web site at http://www.crcpress.com

INTRODUCTION

The discovery, by medical researchers, that the mineral phase of bones and teeth consisted of calcium phosphate salts, has been the start of interest in these salts as a material to enhance healing of bony fractures. This means that calcium phosphate has been in the picture for almost a century. Nevertheless, ceramics of calcium phosphate salts are considered to be "new" materials. The word "new" obviously refers neither to the calcium phosphate salts themselves, nor to their potential importance for bone repair, but to the fact that only at the end of the 1960s the preparation of solids from calcium phosphate powders was achieved. The technique used most often is sintering.

Less than 15 years after the first publication appeared on preparing solid materials of calcium phosphate, this new class of bioceramics is on the verge of being widely applied in the clinic. Several producers have already put their products on the market.

Still, the material has several biomechanical limitations, while some other aspects, like rate of biodegradation, cannot be quantitatively predicted yet. In this book, a number of authors have tried to put together the current knowledge related to calcium phosphate bioceramics.

The first chapters are fundamental, in that the physical chemistry of calcium phosphate salts is discussed, along with mineralization (with emphasis on teeth) and remodelling of mineralized tissues. Thereafter follows a treatment of the influence of implants on surrounding hard tissues. This topic is followed by a chapter on preparation methods and biomechanical and biological aspects of calcium phosphate implants. In conclusion, two chapters are devoted to (possible) dental and medical applications. It is hoped that basic researchers can use the book in their efforts to improve this promising class of materials further, and that clinicians are inspired to define further possibilities and — at least as important — limitations.

THE EDITOR

Klaas de Groot, Ph.D., is Professor of Biomaterials in the Schools of Dentistry and Medicine at the Free University of Amsterdam, The Netherlands. He is head of the Department of Biomaterials. He received his Ph.D. in Physical Chemistry in 1968 from the University of Groningen, The Netherlands. His thesis was based on research carried out at the University of North Carolina, Department of Chemistry. During the years 1968 to 1971, he worked as research associate at various Dutch universities. Since 1971, he has been on the faculty of the Free University, first as Associate Professor of Biomaterials and, since 1980, as full professor. In 1972, he spent a sabbatical half-year at the UCLA Bone Research Laboratory, and in 1977, a summer course as visiting associate professor at the University of Utah. Professor de Groot is a member of several professional societies, among which include the IADR, the Society of Biomaterials, and the European Society of Biomaterials. During the years 1976 to 1980, he was council member of the latter society.

He has been recipient of many research grants and contracts from Dutch funding agencies and collaborated closely with several indutries. He has been Dean of the Dental School of the Free University, and Chairman of its Research Committee. Currently, he is Vice-Dean.

Professor de Groot has authored or co-authored more than 100 scientific communications in the field of biomaterials. His major research interests include bioceramics of calcium phosphate.

CONTRIBUTORS

Klaas de Groot, Ph.D.
Professor of Materials Science
Department of Biomaterials
Schools of Dentistry and Medicine
Free University
Amsterdam, The Netherlands

F. C. M. Driessens, Ph.D.
Professor of Dental Materials and Technology
Dental School
Catholic University of Nijmegen
Nijmegen, The Netherlands

Louw Feenstra, M.D.
Professor of ENT
Department of ENT
Academic Hospital
Free University
Amsterdam, The Netherlands

Peter Griss, M.D., P.D.
Lindenhof Orthopaedic Hospital
Faculty for Clinical Medicine at Mannheim
University of Heidelberg
Mannheim, West Germany

Gunther Heimke, Ph.D.
Friedrichsfeld GmbH
Steinzeug- und Kunststoffwerke
Mannheim, West Germany

J. C. Maltha, Ph.D.
Department of Oral Histology
University of Nijmegen
Nijmegen, The Netherlands

P. J. van Mullem, Ph.D.
Department of Oral Histology
University of Nijmegen
Nijmegen, The Netherlands

J. H. M. Woltgens, Ph.D., D.M.D.
Department of Biomaterials
Schools of Dentistry and Medicine
Free University
Amsterdam, The Netherlands

TABLE OF CONTENTS

Chapter 1

FORMATION AND STABILITY OF CALCIUM PHOSPHATES IN RELATION TO THE PHASE COMPOSITION OF THE MINERAL IN CALCIFIED TISSUES

F. C. M. Driessens

TABLE OF CONTENTS

I. INTRODUCTION

Choices must be made on several levels of magnification for the design of materials for implantation. On the molecular level, it is important to choose either a material that is indifferent under physiological conditions or a material that resembles the tissue for which it is substituted. Calcium phosphates belong to the second category. They are used mainly as substitutes for or improvement of calcified tissues such as bone and dental hard tissues. The main components of the mineral particles in these tissues are calcium and phosphate.

For the preparation of calcium phosphates for the purpose of implantation, knowledge about the formation and stability of calcium phosphates in general is a prerequisite. This chapter is partly intended to convey knowledge about these aspects of calcium phosphates. In this context it is also important to pay attention to the degree of resemblance that can be obtained between synthetic calcium phosphates and the biominerals of the calcified tissues. Therefore, it is necessary to include in this chapter the different concepts published about the nature of these biominerals, either in bone, dentin, or tooth enamel.

In the following chapter, attention will focus on the biological factors that determine the stability of the above-mentioned biominerals under physiological and eventually pathological conditions. This will be done for tooth enamel in Chapter 2 and for bone and bone mineral in Chapter 3. In order to relate the biological stability of the biominerals with these concepts published concerning their nature, an estimate about the physicochemical factors that determine the stability of these biominerals is also presented here.

II. FORMATION OF CALCIUM PHOSPHATES

Polycrystalline calcium phosphates can be obtained in the form of a loose powder by precipitation from aqueous solution. These precipitates can contain different amounts of H^+ and OH^- ions as well as water molecules, depending on the experimental conditions. Through hydrothermal techniques, the temperature can be raised above 100°C and in practice up to about 1000°C. The preparations can also contain "foreign ions" when the precipitations are carried out in their presence in the aqueous solution.

A way to obtain polycrystalline calcium phosphates in the form of more or less dense bodies is by applying ceramic techniques. In this way, the porosity can be lowered either by "hotpressing" or by liquid sintering. Also, the calcium phosphates can be doped with "foreign ions" in these techniques. Both ways of synthesis are relevant to the production of calcium phosphates for the purpose of implantation and, therefore, will be dealt with in this section.

A. Formation of Calcium Phosphates by Precipitation From Aqueous Solutions

Early studies used the addition of Ca^{2+}-containing aqueous solutions to phosphate-containing aqueous solutions or the reverse to obtain precipitates of calcium phosphates. Cameron and Hurst[22] have mentioned the existence of dicalcium phosphate (DCP), dicalcium phosphate dihydrate (DCPD), tricalcium phosphate (TCP), and apatite, of which all but the latter were characterized by their chemical composition at that time, whereas apatite was known as a mineral. Bassett[10] specified the apatite being in equilibrium with aqueous solutions erroneously as oxyapatite, but found firm evidence for metastable equilibria, in which $CaHPO_4$ (DCP) and $CaHPO_4 \cdot 2\ H_2O$ (DCPD) could exist as solid phases, the latter forming preferentially under 36°C.[11] The first author to present the right formula for hydroxyapatite $Ca_{10}(PO_4)_6(OH)_2$ probably was Forster.[58] Schleede, Schmidt, and Kindt[148] were able to prepare chemically pure hydroxyapatite OHA by precipitation. For this purpose, they subjected either an unstable precipitate of defective hydroxyapatite $Ca_9(PO_4)_5(HPO_4)(OH)$ or a high-temperature product of tetracalcium phosphate $Ca_4(PO_4)_2O$ to hydrolysis at temperatures up to 90°C.

The preparation of pure octocalcium phosphate (OCP) was first carried out by Newesely.[123] The exact number of water molecules in its formula $Ca_8(HPO_4)_2(PO_4)_4 \cdot 5H_2O$ was elucidated by the determination of its crystal structure.[19] According to Newesely,[123] OCP is formed in a pure form, when $CaHPO_4$ is hydrolyzed in a solution buffered somewhere between pH 5.8 and 7.1. The temperature must be kept below 65°C at pH 5.8 and below 38°C at pH 7.1. Direct precipitation is also possible at pH 6.5 from acetate-phosphate buffers by the addition of calcium nitrate solutions.

Sanfourche[142] found that under certain conditions, the first precipitate was neither one of the above-mentioned crystalline compounds. It was "gelatinous" and it could take days before it was transformed into one of these crystalline compounds. Amorphous calcium phosphate (ACP) has been characterized by several methods.[161] It contains HPO_4^{2-} ions and its Ca/P ratio was found to be 1.36 ± 0.02, initially. Upon washing it changed to 1.47 ± 0.03.[163] However, unwashed samples had a constant Ca/P ratio of 1.42 ± 0.02 during suspension over several hours (see Table 2).[163] This indicates that ACP is primarily a very finely dispersed whitlockite which can absorb either HPO_4^{2-} or Ca^{2+} ions. The structure of the mineral whitlockite was studied by Frondel.[60] It can also exist without foreign ions and then has the composition $Ca_{10}(HPO_4)(PO_4)_6$.[147] The conclusion that ACP is a very finely dispersed whitlockite is enforced by the fact that Meyer and Eanes[110] found that the ionic product of whitlockite is constant over ACP suspensions.

A compilation of the calcium phosphates pertinent to aqueous systems is given in Table 1, together with the space group of their crystal structure and the most reliable value of their solubility product.

The occurrence of metastable states during the precipitation and formation of calcium phosphates has already been mentioned. Even among more recent studies there is not much agreement about the nature of the first precipitate and the consecutive transitions into other calcium phosphates, although physical methods like monitoring of pH[188] and X-ray diffraction analysis[168] have been applied in this field since their development. The latter authors postulate the formation of defective hydroxyapatites which can vary in composition from $Ca_9(PO_4)_6(H_20)_2$ to $Ca_{10}(PO_4)_6(OH)_2$. Sanfourche[142] observed that it could take days before such compounds were formed from the original "gelatinous" precipitate which apparently was ACP. Fouretier[59] found that the first precipitate was a mixture of DCPD and ACP, which hydrolyzed to a "hydrated" apatite. Arnold[4] proposed that OCP and OHA form a continuous series of solid solutions so that no two preparations with composition lying between them

Table 1
PERTINENT CALCIUM PHOSPHATES RELEVANT TO AQUEOUS SYSTEMS, THEIR FORMULA, STRUCTURE AND NEGATIVE LOGARITHM OF THE SOLUBILITY PRODUCT pK_{sp}

Ca/P	Notation	Formula	Space group	pK_{sp}	Mineral same
1	DCP	$CaHPO_4$	$P\bar{1}$	6.90	Monetite
1	DCPD	$CaHPO_4 \cdot 2H_2O$	C2/c	6.59	Brushite
1.33	OCP	$Ca_8(HPO_4)_2(PO_4)_4 \cdot 5H_2O$	$P\bar{1}$	68.6[a]	—
1.43	WH	$Ca_{10}(HPO_4)(PO_4)_6$	R3c	81.7[a]	Whitlockite
1.67	OHA	$Ca_{10}(PO_4)_6(OH)_2$	$P6_3/m$	117.2	Hydroxyapatite
1.50	DOHA	$Ca_9(HPO_4)(PO_4)_5(OH)$	$P6_3/m$	85.1[a]	Defective hydroxy-apatite

[a] Estimate.

would be likely to be identical. Strates, Neuman, and Levinkes[158] found that DCPD is the first precipitate and that this hydrolyzes into a defective hydroxyapatite. Brown, Smith, Lehr, and Frazier[20] postulated that OCP and OHA do not form solid solutions, but form instead epitactical and topotactical mixtures with a sandwich structure. Eanes, Gillesen, and Posner[46] first found an amorphous calcium phosphate which hydrolyzes in about 5 hr into a defective apatite. Newesely[126] proved that the type of first precipitate depends largely on the degree of supersaturation. At low supersaturation, DCPD was formed which, above pH 6.3, hydrolyzed into OCP. At high supersaturation an amorphous product was formed directly from aqueous solution. Combining this information, one comes to the scheme of reactions given in Figure 1. Most experiments mentioned here were carried out in the range $5<pH<8$. It must be kept in mind that the transformations are pH-sensitive[126] which may explain, together with the varying degrees of supersaturation, the apparent differences found between the various authors.

Meyer and Eanes[111] found that the aqueous solutions in equilibrium with the crystalline phase first formed from OCP show up the constant solubility product of OCP. However, at the end of the amorphous-to-crystalline transformation, not the solubility product of OCP, but that of TCP appeared to be constant. In agreement with this Heughebaert and Montel[76] found that the composition of the crystalline phase right after completion of its formation from ACP agrees with the formula $Ca_9(HPO_4)_x(PO_4)_{6-x}(OH)_x$. This compound at x=1 will be called defective hydroxyapatite (DOHA). It is not clear yet whether preparations having a composition between that of DOHA and OHA are solid solutions or mixtures of these two apatites.

From the reaction scheme of Figure 1, it is seen that OHA is the end product. This appears to be so in the range pH > 4. Elevation of the temperature increases the rate with which the end product is formed. Therefore, the most common method of preparation of OHA is by precipitation from a boiling aqueous solution.[178] Hydrothermal syntheses at higher temperatures and pressures have also been carried out.[3,140,154] The main effect is the increase of the average particle size.

1. Incorporation of "Foreign" Ions: Fluoride

Preparation of calcium phosphates in the presence of certain "foreign" ions results in the partial incorporation of these ions in the crystals of the precipitate. In this way, Moreno, Kresak, and Zahradnik[116] were able to prepare solid solutions of OHA and fluorapatite FA of the formula

$$Ca_5(PO_4)_3OH_{1-x}F_x \qquad (1)$$

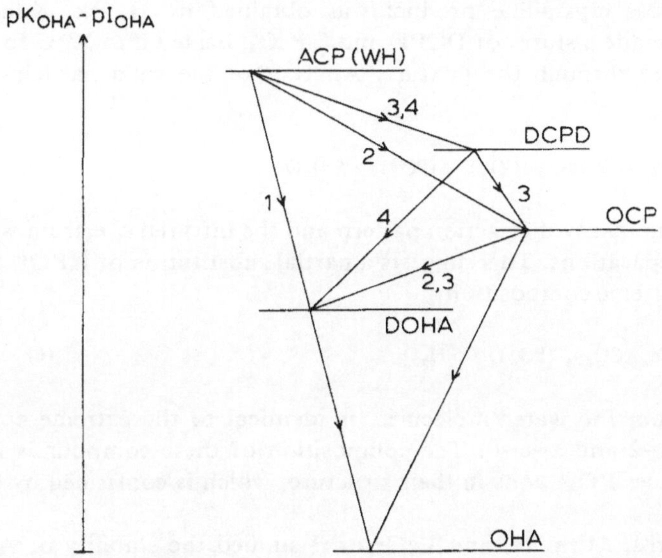

FIGURE 1. Pathways of maturing of calcium phosphate precipitates. (From Driessens, F. C. M., *Bull. Soc. Chim. Belg.*, 89, 663, 1980. With permission.)

by mixing an ammonium phosphate and ammonium fluoride-containing solution with a calcium nitrate solution at a certain rate at boiling temperatures. CO_2 was excluded from the reaction vessel. Care must be taken in this technique to avoid the formation of CaF_2 as an impurity phase due to its limited solubility.

It appears that the presence of F^- ions in the aqueous solution promotes the formation of the apatite phase.[7,8,47,124] It curtails the metastable states. Hydrolysis of acidic calcium phosphates such as DCPD and OCP into OHA is also accelerated by the presence of some F^- ions in the aqueous solutions.[43,44,123]

2. Incorporation of "Foreign" Ions: Carbonate

The CO_3^{2-} ions also interfere with calcium phosphates. Labarthe, Bonel, and Montel[97] succeeded in obtaining carbonated apatites by co-precipitation of calcium carbonate and phosphate. X-ray analysis of the products revealed their apatitic nature, whereas chemical analysis showed that their structure formula is probably

$$Ca_{10-x+u}\,(PO_4)_{6-x}\,(CO_3)_x\,(OH)_{2-x+2u} \qquad (2)$$

with $0 \leqslant x \leqslant 2$ and $0 \leqslant u \leqslant \frac{1}{2}x$.[98] An alternative formula for the same compounds was proposed by Bonel, Labarthe, and Vignoles:[15]

$$Ca_{10-x+u}\,(PO_4)_{6-x}\,(CO_3)_x\,(H_2O)_z\,(OH)_{2-x+2u} \qquad (3)$$

Through the equilibrium

$$H_2O + PO_4^{3-} \rightleftharpoons OH^- + HPO_4^{2-} \qquad (4)$$

which was proven to play a role in defective apatites by infrared spectroscopy.[62] Equation 3 might account for HPO_4^{2-} ions and for additional OH^- groups within these apatites. The lattice parameters of such preparations are not yet well established and apparently depend on the way of precipitation.[99]

A probably less crystalline product was obtained by Hayek, Konetschny, and Schnell.[75] They made a slurry of DCPD and $CaCO_3$, heated it to 37°C for several days while air bubbled through the mixture. Analysis of the solid particles revealed the composition

$$Ca_8\,(HPO_4)_{0.8}\,(CO_3)_{1.2}\,(PO_4)_4 \cdot 5\,H_2O \tag{5}$$

whereas the vague X-ray diffraction pattern and the infrared spectrum were similar to that of OCP preparations. This suggests a partial substitution of HPO^{2-} by $CO_3{}^{2-}$ ions leading to the extreme composition

$$Ca_8\,(CO_3)_2\,(PO_4)_4 \cdot 5\,H_2O \tag{6}$$

which, apart from the water molecules, is identical to the extreme composition of Equation 3 at x=2 and z=u=0. The composition of these compounds indicates that $CO_3{}^{2-}$ ions replaces $PO_4{}^{3-}$ ions in their structure, which is confirmed by their infrared spectrum.[15]

LeGeros, Shirra, Miravite, and LeGeros[103] studied the stability of ACP in $CO_3{}^{2-}$-containing aqueous solutions and found that the crystallization was retarded. Although it is not clear in which parts of the pathways of Figure 1 this happens, it might well be in the first crystallization to OCP, as Newesely[124] had suggested that carbonate reacts with OCP.

Eysel and Roy[51] carried out a reaction between phosphate-containing aqueous solutions and $CaCO_3$ in the form of aragonite under hydrothermal conditions in the range of 260° to 400°C. They found an intergrowth and overgrowth of the aragonite crystals with an apatite-like calcium phosphate. They[141] also claimed that in this way porous, but coherent apatite bodies could be made from coral which might be suitable for implantation.

3. Incorporation of "Foreign" Ions: Sodium and Carbonate

Particular attention must be paid to the combined effect of Na^+ and $CO_3{}^{2-}$ incorporation into calcium phosphates. Early reports on the preparation of such compounds do not mention their alkali content.[94,136] Newesely[125] prepared a number of apatites containing sodium and carbonate by hydrothermal synthesis at 400°C. Newesely mentioned that the samples contained between 0.5 and 1.1% sodium without giving the exact composition of each individual product.

Simpson[151] prepared carbonated apatites containing up to 2.2% sodium by reaction of $CaCO_3$ with Na^+ and phosphate-containing aqueous solutions. The CO_3 content was about 7.5%. Their X-ray diffraction pattern was that of well crystalline apatites, when prepared at pH 7. Later Simpson[152] showed that there was an inverse relation between the carbonate content of the precipitate and the $PO_4{}^{3-}$ content of the aqueous solution during precipitation.

Zapanta-LeGeros[193] prepared well crystalline, sodium- and carbonate-containing apatites by dropwise addition of a calcium acetate solution to a boiling aqueous solution of Na_2HPO_4 and $NaHCO_3$. Carbonate probably substitutes for phosphate in their structure because there was an inverse relation between the carbonate and the phosphate content.[101] The sodium content was proportional to the carbonate content.[100]

Vignoles, Bonel, and Montel[184] have repeated the preparations of these apatites according to LeGeros and reported that their composition agrees with the formula

$$Ca_{10-x}\,Na_{2/3\,x}\,(PO_4)_{6-x}\,(CO_3)_x\,(H_2O)_y\,(OH)_{2-1/3\,x} \tag{7}$$

whereby $0 \leqslant x \leqslant 3$ and $0 \leqslant y \leqslant x$. According to this formula the Ca/P ratio can vary from 1.67 to 2.33

4. Incorporation of "Foreign" Ions: Potassium and Carbonate

The combination of potassium with carbonate has similar effects on calcium phosphates. Simpson[151] reacted potassium phosphate-containing aqueous solutions with calcite. Below pH 6.1 only DCPD formed. Above pH 6.1 a poorly crystalline apatite formed which contained about 3% CO_3 and 0.5% potassium. Above pH 8 better crystalline apatites were formed. Later, Simpson[153] reported on the formation of potassium- and carbonate-containing apatites by evaporation of water from a solution containing Ca-EDTA and potassium phosphate 100°C between pH 6 and 14. These products contained up to 1% CO_3 and 1% potassium.

LeGeros[100] prepared potassium- and carbonate-containing apatites from a boiling solution of potassium phosphate and potassium carbonate by adding dropwise a solution of calcium acetate. From the reported chemical composition, the following approximate formula can be calculated:

$$Ca_{10-\frac{1}{2}x}\,K_{\frac{1}{2}x}\,(PO_4)_{6-\frac{1}{2}x}\,(CO_3)_x(OH)_{2-\frac{1}{2}x} \qquad (8)$$

The products were contained in the range of $1 \leqslant x \leqslant 3$.

5. Incorporation of "Foreign" Ions: Magnesium

The presence of Mg^{2+} ions in the aqueous solution stabilizes the structure of whitlockite and causes its formation in preference of other calcium phosphates under conditions of precipitation.[74] Boulet, Marier, and Rose[17] produced evidence from precipitation studies that Mg^{2+} ions are incorporated in the "gelatinous" precipitate of ACP. Bachra, Trautz, and Simon[7] determined a so-called precipitation diagram for the system "calcium-carbonate-phosphate" and found that in the presence of Mg^{2+} ions, the ACP phase field expanded at the expense of the DOHA field. In another paper they reported that the Mg^{2+} ion stabilized the ACP precipitate and disturbed the crystallization of the apatite phase. Termine and Posner[161] reported that at a given degree of supersaturation the presence of Mg^{2+} ions enhanced ACP formation. Rowles[139] prepared Mg-containing whitlockite by precipitation and found that its formula was

$$Ca_9\,(Ca_{1-x}\,Mg_x)(HPO_4)\,(PO_4)_6 \qquad (9)$$

for preparations below 100°C. Gopal, Calvo, Ito, and Sabine[63] made such preparations under hydrothermal conditions and found the same formula for their products prepared in the range of 700 to 1000°C.

The presence of Mg^{2+} ions can retard the transformation of ACP (finely dispersed whitlockite) into other calcium phosphates. Termine, Peckauskas, and Posner[162] determined the time that elapsed during the solution-mediated transformation of ACP into DOHA. They found that addition of small amounts of Mg^{2+} ions increased this period considerably. Brown, Smith, Lehr, and Frazier[20] reported that Mg^{2+} ions in a concentration as low as 0.001 mol l^{-1} completely blocked the hydrolysis of OCP into DOHA. Boulet, Marier, and Rose[17] found that Mg^{2+} ions had no effect on the formation of DCPD in precipitation studies from slightly acidic solutions. Neuman and Mulryan[122] carried out exchange studies on DOHA slurries with Mg^{2+}-containing aqueous solutions and found that the Mg^{2+} ions were not incorporated in the solid apatite.

6. Incorporation of Other "Foreign" Ions

Because of their biological importance the effect of the "foreign ions" mentioned here has been investigated thoroughly. Other "foreign ions" may have similar effects like Sr^{2+},[174] Pb^{2+}, Zn^{2+}, Mn^{2+}, Fe^{3+}, and Cd^{2+}.[23-25,56,120,192] They mainly affect the apatite structure, but are not likely to be biologically advantageous for making calcium phosphates for implantation purposes.

B. Formation of Calcium Phosphates By Solid-State Reaction

Solid state reactions at temperatures beyond about 700°C ultimately result in one calcium phosphate or a mixture of more calcium phosphates, of which the composition depends only on the proportioning of nonvolatile components, the partial pressure of the volatile components, the temperature, and the total pressure. For this reason it is not necessary to name the exact ingredients from which the different authors obtained the several calcium phosphates. As calcium phosphates with a ratio $Ca/P < 1$ are easily soluble and hydrolyze very quickly, they are not suitable for implantation purposes and therefore will not be considered here.

In the range $Ca/P \geqslant 1$ the phases CPP or calcium pyrophosphate $Ca_2P_2O_7$, TCP or tricalcium phosphate $Ca_3(PO_4)_2$ and tetracalcium phosphate $Ca_4(PO_4)_2O$ are known to be formed from the melt. The first two phases occur in different crystallographic forms as a function of temperature, as will be shown in Section III. Tetracalcium phosphate is not stable in air below about 1420°C, but reacts with water vapor according to:[167]

$$3\,Ca_4\,(PO_4)_2\,O + H_2O \rightarrow Ca_{10}\,(PO_4)_6(OH)_2 + 2\,CaO \qquad (10)$$

Although Tromel[167] rightly claimed that pure oxyapatite $Ca_{10}(PO_4)_6O$ does not exist, Riboud,[131] Seuter,[150] and Verbeeck, Heyligers, Driessens, and Schaeken[179] produced evidence that "hydroxyapatites" obtained by solid-state reaction at high temperatures are in fact oxyhydroxyapatites of the formula

$$Ca_{10}\,(PO_4)_6\,(OH)_{2-2x}\,O_x\,V(OH)_x \qquad (11)$$

in which V(OH) means a vacancy in the OH sublattice. The value of x depends on the temperature and the partial water vapor pressure.

Welch and Gutt[187] have shown that α-TCP as well as β-TCP, being the high and low temperature form of TCP, both show another type of deviation from stoichiometry. They both seem to be stable in a certain range $Ca/P < 3/2$. These authors made their investigations at temperatures beyond 1000°C. Recently, Schaeken, Driessens, and Verbeeck[147] found that "β-TCP" prepared at about 900°C under moist CO_2 has the stoichiometry of whitlockite and is stable only in a narrow range: $Ca/P = 1.48 \pm 0.01$. Whether these preparations contain water is still under investigation. In this context it is important to note that the transformation α-TCP$\leftrightarrow\beta$-TCP depends strongly on the partial water vapor pressure.[18] It is also noteworthy that these authors[18] observed that extra CaO was necessary to form α-TCP out of β-TCP. Although these data show that one might doubt whether the nature of "β-TCP" in fact is that of whitlockite, the designation β-TCP will be maintained in this paper to preparations of this phase at high temperature.

1. Incorporation of Halides

Wallaeys[186] demonstrated that fluorapatite $Ca_{10}(PO_4)_6F_2$ and chlorapatite $Ca_{10}(PO_4)_6Cl_2$ were formed when calcium phosphates were heated with CaF_2 and $CaCl_2$, respectively. This was confirmed by Montel[113] for fluorapatite. Montel also showed that in the absence of water vapor, the excess of phosphorus and fluoride volatilizes as POF_3. Later, Prener[132] showed that these apatites had deviations from stoichiometry according to the formula

$$Ca_{10-x}(PO_4)_6F_{2-2x} \quad (12)$$

due to the fact that they tend to be in equilibrium with CaF_2 vapor.

Schaeken, Verbeeck, Driessens, and Thun[145] intended to prepare solid solutions of hydroxyapatite and fluorapatite of the formula

$$Ca_{10}(PO_4)_6(OH)_{2-2x}F_{2x} \quad (13)$$

by heating mixtures of $CaHPO_4$, $CaCO_3$, and CaF_2 under moist air. Their samples lost some fluoride during this heating and they propose a formula

$$Ca_{10}(PO_4)_6(OH)_{2-2x-2y}F_{2x}O_yV(OH)_y \quad (14)$$

which involves deviations from stoichiometry of the type in Equation 11.[176]

Wallaeys and Chaudron[185] prepared solid solutions of fluorapatite and chlorapatite of the formula

$$Ca_{10}(PO_4)_6Cl_{2-2x}F_{2x} \quad (15)$$

by heating mixtures of chlorapatite and CaF_2. Heating results in the formation of $CaCl_2$ as a byproduct which shows that fluorapatite is more stable than chlorapatite.

From the work of McConnell,[108] Hägele and Machatschki,[70] and Klement,[90] it follows that the phosphorus atoms in fluorapatite can be replaced partially or totally by other atoms. Charge compensation can be achieved in various ways:

2 P^{5+} replaced by Si^{4+} and S^{6+}

Ca^{2+} and P^{5+} replaced by Me^{3+} and Si^{4+}, in which Me^{3+} = rare earth

Ca^{2+} and P^{5+} replaced by Me^{+} and S^{6+}, in which Me^{+} = Na^{+}

Klement and Haselbeck[91] described mixed apatites of the formula

$$Ca_{10-x}Me_x(PO_4)_6Cl_2 \quad (16)$$

where Me stands for Zn and Cu (x up to 4) and for Co (x up to 3). A compound of the formula $Ca_8Na_2(PO_4)_6$ has been reported by Harth.[73] Kreidler and Hummel[93] showed that fluorapatites similar to Equation 16 can be formed for nickel and magnesium (up to x=0.8).

2. Incorporation of Silicate

Nurse, Welch, and Gutt[127] investigated the system $Ca_2SiO_4-Ca_3(PO_4)_2$. They found that the incorporation of small amounts of Ca_2SiO_4 in TCP stabilized the α-TCP structure down to lower temperatures. In this system, silicocarnotite is found as a separate phase. Its composition varies somewhat around the formula $Ca_5(PO_4)_2(SiO_4)$ and it is stable up to 1450°C. Its structure is somewhat related to that of hydroxyapatite.[31] Nurse, Welch, and Gutt[127] also formed a new compound having the formula $Ca_7(PO_4)_2(SiO_4)_2$ which is stable up to 1125°C. They confirmed the existence of a very high temperature form of TCP, the so-called $\bar{\alpha}$-TCP form which is isostructural with Ca_2SiO_4 and which forms a continuous series of solid solutions at temperatures greater than 1440°C.[9] Such solid solution at the composition $Ca_9(PO_4)_2(SiO_4)_3$ seems to be stable down to 500°C and may be lower. Their results have been confirmed largely by Fix, Heymann, and Heinke.[55]

3. Incorporation of Sulfide and Sulfate

Trombe and Montel[172] reported the formation of sulfoapatite $Ca_{10}(PO_4)_6S$ by heating hydroxyapatite in a dry helium atmosphere containing sulfur vapor at a pressure of at least 2 mmHg at 1000°C. Heating of the sulfoapatite at a temperature between 750 and 1000°C resulted in the formation of a β-TCP type compound containing sulfate.[173]

4. Incorporation of Carbonate

Well crystalline, single-phase carbonate-containing calcium phosphates can be obtained by solid-state reaction at high temperatures in a CO_2-containing atmosphere. By heating OHA in CO_2, Wallaeys and Chaudron[185] obtained a partially carbonated hydroxyapatite with the formula

$$Ca_{10}(PO_4)_6(OH)_{0.8}(CO_3)_{0.6} \qquad (17)$$

Reheating of this product in air yielded OHA so that the carbonation is reversible. Bonel[14] was able to prepare several compositions between OHA and carbonato-apatite $Ca_{10}(PO_4)_6CO_3$ of the general formula

$$Ca_{10}(PO_4)_6(OH)_{2-2x}(CO_3)_x \qquad 0 \leq x \leq 1 \qquad (18)$$

by application of dry CO_2 and varying reaction times. The lattice parameters of these apatites were not a linear function of x. According to Bonel,[14] the carbonato-apatite is not hexagonal at room temperature, but has a slightly distorted structure. The space group and the temperature of transition into hexagonal apatite have not been determined as yet.

Recently, Schaeken, Driessens, and Verbeeck[147] succeeded in preparing carbonated apatites with Ca/P ratios larger than 1.67. The composition is close to the general formula

$$Ca_{10-x}(HPO_4)_x(PO_4)_{6-x}(CO_3)_{1-\frac{1}{2}x} \qquad (19)$$

when prepared at 870°C in an atmosphere of CO_2 containing a small partial water vapor pressure. Under these conditions they are stable, single-phase products in the range $0 \leq x \leq 0.5$ so that the Ca/P ratio can vary from 1.67 to 1.76. No products could be synthesized having a Ca/P ratio smaller than 1.67 under these condtions.

In all these preparations, the apatite structure shows an increased a axis and a decreased c axis in comparison to OHA.[14,39,102] Schaeken, Driessens, and Verbeeck[147] proved that carbonate is not incorporated in the whitlockite phase at high temperatures.

On the basis of infrared absorption spectra Bonel[14] concluded that all CO_3^{2-} ions in those preparations having the formula in Equation 18 are contained in OH positions, thereby leaving half of the OH positions vacant. Driessens, Schaeken, Heyligers, and Verbeeck[39] confirmed this result for carbonato-apatite. By measuring infrared spectra, chemical compositions, lattice parameters, and densities they found that the apatite of the composition

$$Ca_{10}(PO_4)_{5.65}(CO_3)_{1.09}(OH)_{0.27} \qquad (20)$$

approaches closely to the structure

$$Ca_{9.84}(PO_4)_{5.56}(CO_3)_{0.44}(CO_3)_{0.63}(OH)_{0.86} \qquad (21)$$

so that 2/5 of the CO_3^{2-} ions are located on PO_4^{3-} sites and 3/5 on OH sites, whereas there are vacancies in the Ca and the OH sublattice.

5. Incorporation of Sodium and Carbonate

Schaeken, Driessens, and Verbeeck[147] heated mixtures of $CaHPO_4$, $CaCO_3$, and Na_2CO_3 at temperatures between 800 and 1000° C in CO_2 having a low partial water vapor pressure. It could take several weeks before the reaction was complete. They found that sodium is incorporated in the whitlockite phase and the apatite phase.

The single-phase products having the whitlockite structure correspond approximately with the formula

$$Ca_{20}\,Na_{2X}\,(P_2O_7)_{1-X}\,(PO_4)_{12+2X} \qquad (22)$$

These products are stable under the conditions of preparation in the range $0 \leqslant x \leqslant 1$. Their experimental Ca/P ratio is 1.43 ± 0.02 (theor. 1.429).

The composition of the single-phase products having the apatite structure complies with the formula

$$Ca_{10-X}\,Na_X\,(PO_4)_{6-X}\,(CO_3)_{1+X} \qquad (23)$$

within the limits of experimental error. Under the conditions of preparation, these apatites are stable in the range $0 \leqslant x \leqslant 1.5$ so that their Ca/P ratio can vary from 1.67 to 1.89. According to Driessens, Schaeken, Heyligers, and Verbeeck[39] the lattice parameter a of these compounds decreases with x, whereas the c axis increases with x. Their average refractive index diminishes from 1.650 at x=0 to 1.634 at x=1.5.

The infrared spectrum of the apatite of Equation 23 at x=1.5 was such that according to Elliott's[48] assignment of bands, all of its CO_3^{2-} ions would be contained in PO_4 positions. However, measurements of the crystal lattice parameters and the densities by Driessens, Schaeken, Heyligers, and Verbeeck[39] showed that the structure is given by the formula

$$Ca_{8.5}\,Na_{1.5}\,[(PO_4)_{4.5}\,(CO_3)_{1.5}]\,(CO_3) \qquad (24)$$

so that 40% of the CO_3^{2-} ions are in OH positions.

6. Incorporation of Magnesium

Whitlockite is the only calcium phosphate which incorporates Mg^{2+} ions. Various formulas have been proposed for these whitlockites. Tovborg Jensen and Rowles[165] proposed

$$Ca_{21-X}\,Mg_X\,(PO_4)_{14} \qquad (25)$$

with $0 \leqslant x \leqslant 2$, which was derived from the phase composition of samples prepared by solid-state reaction at high temperatures. On the basis of literature data about the composition and the content of the elementary cell Keppler[85] concluded that the formula should be

$$Ca_9\,(Ca_{1-X}\,Mg_X)\,(HPO_4)\,(PO_4)_6 \qquad (26)$$

whereby the unit cell contains two formula units. Keppler[86] also showed that the dimensions of the elementary cell at x=1 are only slightly smaller than those at x=0.

Ito[80] prepared magnesium-containing whitlockites by heating the appropriate ingredients at 1120°C in air. Ito proposed the formula

$$Ca_{9-x}Mg_x(PO_4)_6 \quad (27)$$

and claimed that the products were single-phase whitlockites up to x=2.

In a thorough study Gopal, Calvo, Ito, and Sabine[63] showed that the composition of Mg-whitlockite grown by hydrothermal synthesis agrees with Equation 26 at x=1. They also determined the crystal structure. Rowles[139] proved that this formula also applies to preparations obtained from aqueous solutions under 100°C.

7. Miscellaneous Incorporations: Rhenanite

Ando[1] determined the mutual solubility of $Ca_3(PO_4)_2$ and $Mg_3(PO_4)_2$ by heating mixtures in air and arrived at the same formula (Equation 27) as Ito,[80] whereby the maximum value of x appeared to be 1. Ando showed that magnesium incorporations stabilized the β-TCP structure from about 1180°C up to 1500°C. Ando and Matsumo[2] investigated the system $Ca_3(PO_4)_2$–$CaNaPO_4$ and confirmed that compounds like that given by Equation 22 could be formed. They also claimed that the high temperature form of rhenanite $CaNaPO_4$ could dissolve large amounts of $Ca_3(PO_4)_2$ at high temperatures.

Schaeken, Driessens, and Verbeeck[147] found that rhenanite has an extended phase field by heating under CO_2. Single-phase products had the formula

$$Ca_{1-x}Na_{1+x}(PO_4)_{1-x}(CO_3)_x \quad (28)$$

Similarly, Klement and Steckenreiter[92] found that compounds of the formula

$$Ca_{1-x}Na_{1+x}(PO_4)_{1-x}(SO_4)_x \quad (29)$$

could be formed by solid-state reaction.

Kreidler and Hummel[93] investigated the system $Ca_3(PO_4)_2$–$Zn_3(PO_4)_2$ and found that the effect of zinc incorporation into β-TCP was the same as that of magnesium incorporation reported by Ando.[1] About 11% of the calcium ions could be replaced by zinc and the latter element stabilized the β-TCP structure up to about 1500°C.

McCauley, Hummel, and Hoffmann[107] confirmed the mutual solubility of $Ca_3(PO_4)_2$ and $Mg_3(PO_4)_2$ reported by Ando.[1] They also found that about 10% of the calcium ions in $Ca_2P_2O_7$ could be replaced by Mg^{2+} ions.

According to Sarver, Hoffman, and Hummel[144] the high temperature form of $Sr_3(PO_4)_2$ is isostructural with β-TCP and they form a continuous series of solid solutions. The same holds for the hydroxyapatites of calcium and strontium.[77]

8. Solid-State Reactions and Density

In the above-mentioned studies of this section (II.B.) the investigators have mainly focused on the formation and stability of calcium phosphates at high temperatures and not to their density. The following studies using ceramic techniques were especially devoted to the aspect of density of the ceramic calcium phosphate products. This aspect is important insofar as high crystallinity and large material density result in resistance to dissolution and long-lasting stability, while amorphous ultrastructure and porous formation enhance interface activity and bone ingrowth, but also biological degradation of the ceramic implant material.[129]

Normal sintering results in a density for hydroxyapatite of about 90% whereas isostatic pressing and fractional drying of the green body may result in a 97% density.[129] The technique of continuous "hotpressing" applied at 900°C results at once in a density of 97% for hydroxyapatite.[30] De With, van Dijk, Hattu, and Prijs[191] report densities varying from 72 to 96% as a function of sintering temperature between 1100 and 1250°C, when normal sintering after isostatic pressing was carried out for 6 hr. They also showed that a high correlation exists between density and mechanical properties such as Young's modulus, compressive strength, and tensile strength. They reported an average tensile strength of 25 MPa (about 245 kg/cm^2). In other studies, compressive strengths between 1500 and 6000 kg/cm^2 have been reported.[45,159] For brittle materials, compressive strength is mostly about a factor 10 larger than tensile strength.

Schaeken and Driessens[146] have prepared apatites of Equation 23 at x=1.5 by liquid sintering. The proportioning of $CaHPO_4$, $CaCO_3$, and Na_2CO_3 as the starting materials was such that during the sintering at 870°C under CO_2, a small amount of liquid phase was present. The authors measured a density for their specimens equal within the limits of experimental error to the theoretical density of 3.01 g/cm^{-3}. The tensile strength varied from 300 to 500 kg/cm^{-2} and the compressive strength was about a factor 10 higher. The advantages of this are not only the fact that high density is reached within a short time (a few hours) at a relatively low temperature, but also that there is a high degree of similarity of the apatite with the mineral in bone and teeth as will be shown in the next sections.

III. STABILITY OF CALCIUM PHOSPHATES

The stability of calcium phosphates for implantation purposes may be important in two respects. The first is their stability at high temperatures that determines the homogeneity of calcium phosphates made by ceramic techniques. The fact is that certain calcium phosphates are in equilibrium with a liquid phase at certain temperatures and certain systems can be used for the design of experiments in which so-called liquid sintering is applied to obtain dense calcium phosphate bodies without applying any "hotpressing", but simply by sintering at ordinary atmospheric pressure.

Knowledge about the stability of calcium phosphates in contact with aqueous solutions at room temperature or body temperature is also important for implantation purposes. It enables the estimation of reactions that may occur between calcium phosphate implants and their physiological environment.

A. Stability of Calcium Phosphates at High Temperatures

1. The Binary System $CaO-P_2O_5$

The CaO-rich side of this system, which is of special concern here, was described first in a reliable way by Trömel[169] and later by Kreidler and Hummel[93] and by Trömel and Fix.[170] Tetracalcium phosphate melts incongruently at 1710°C. However, TCP and calcium pyrophosphate melt congruently, at 1310° and 1350°C, respectively. Welch and Gutt[187] focused on the fact that both α-TCP and β-TCP could incorporate considerable amounts of P_2O_5 in their structure. Similarly, Hill, Faust, and Reynolds[78] showed that the low temperature form of calcium pyrophosphate β-$Ca_2P_2O_7$ also incorporates excess P_2O_5 in its structure.

Most of these studies concentrated on solid-liquid equilibria, but their validity may be limited as far as solid-state equilibria in this system is concerned. For comparison, Schaeken, Driessens, and Verbeeck[147] have produced firm evidence that β-$Ca_3(PO_4)_2$ is unstable at 870°C. Instead, even at very low water vapor pressures, a whitlockite phase was formed at that temperature. It clearly had the theoretical Ca/P ratio of 1.48 according to Equation 26. This makes the stability of β-$Ca_3(PO_4)_2$ questionable at all, although this compound can be formed easily by cooling α-$Ca_3(PO_4)_2$ samples[32] fast enough down to room temperature.

2. The Ternary System $CaO-P_2O_5-H_2O$

This has been investigated partially by Riboud[134] and more completely by Seuter[150] for the CaO-rich side in the temperature range 950 to 1230°C, by thermogravimetry in combination with X-ray diffraction. Seuter found that in the range $0 < P/Ca < 0.66$ and $1 < PH_2O < 1000$ mmHg stable solid phases were CaO, $Ca_3(PO_4)_2Ca_4(PO_4)_2O$, and apatite. The composition of the latter was dependent on temperature and partial water vapor pressure. Its composition is given by the formula

$$Ca_{10}\,(PO_4)_6\,(OH)_{2-2X}\,O_X\,V(OH)_X \qquad (30)$$

in which V(OH) is a vacancy in the OH sublattice. The following equilibrium was found

$$2\,OH^-\,(s) \rightleftharpoons O^2\,(s) + V(OH\,(s) + H_2O(g) \qquad (31)$$

so that the apatite is in fact an oxyhydroxyapatite. In accordance with the preceeding section pure oxyapatite was found to be unstable.

3. Other Ternary Systems

According to Kreidler and Hummel,[93] Zn-containing β-TCP melts incongruently around 1500°C, whereas the compound $CaZn_2(PO_4)_2$ melts incongruently at 1040°C. They form an eutectic solidifying at 1029°C.

Trömel, Fix, and Fritze[171] investigated the system $CaO-P_2O_5-FeO_n$ in the range from 1550 to 1675°C. They found that TCP and/or tetracalcium phosphate could be in equilibrium with liquid. The behavior of the system $CaO-P_2O_5-MnO_n$ in the range from 1510 to 1610°C is very similar.[96]

Additions of either Al_2O_3[155] or SiO_2[55] or both[156] to CaO/P_2O_5 mixtures do not result in eutectics lower than those already occurring in the binary system $CaO-P_2O_5$.

4. Quaternary and More Complex Systems

St. Pierre[156] has published solid-liquid equilibria in a quaternary cross-section through the system $CaO-Al_2O_3-P_2O_5-SiO_2$. The lower temperature at which $Ca_3(PO_4)_2$ could be in equilibrium with a liquid was 1280°C.

Sata[143] published a quasibinary diagram $Ca_3(PO_4)_2-MgSiO_3$ in the system $CaO-MgO-P_2O_5-SiO_2$ which showed that $Ca_3(PO_4)_2$ could be in equilibrium with a liquid at 1246°C and up.

Schaeken, Driessens, and Verbeeck[147] investigated a quasi-binary cross-section of the system $CaO-P_2O_5-CO_2-H_2O-Na_2O$ at 870°C. They showed that an apatite of Equation 23 at x=1.5 was in equilibration with liquid.

A survey of the ranges of temperature in which a calcium phosphate can be in equilibrium with a liquid phase, is given in Table 2. As it is also important to know the solid solubility of foreign compounds into a calcium phosphate, a survey of these is given in Table 3.

5. Conclusion

The lowest temperature at which an apatite phase is in equilibrium with a liquid phase in the known systems under normal atmospheric pressure, is in the system $CaO—P_2O_5—H_2O—CO_2—Na_2O$. According to the knowledge about the subsystem $CaO—Na_2O—CO_2$,[13] the lowest temperature at which apatite might be in equilibrium with a liquid phase is probably 780°C. The according apatite contains Na and CO_3.

The lowest temperature at which $Ca_3(PO_4)_2$ or perhaps whitlockite is in equilibrium with a liquid phase at high temperatures up to now is found in the system

Table 2
EQUILIBRIA BETWEEN A SOLID CALCIUM PHOSPHATE PHASE (Ca/P ≥ 1) AND A LIQUID PHASE AT HIGH TEMPERATURES IN DIFFERENT SYSTEMS

Solid calcium phosphate	System	Temperature range (degress celsius)	Ref.
$Ca_4(PO_4)_2O$	$CaO-P_2O_5$	1578—1720	170
$Ca_3(PO_4)_2$	$CaO-P_2O_5$	1288—1810	170
$Ca_2P_2O_7$	$CaO-P_2O_5$	985—1350	78
$Ca_3(PO_4)_2$	$CaO-P_2O_5-FeO$	1370—1730	128
$Ca_{10}(PO_4)_6F_2$	$CaO-P_2O_5-CaF_2$	1205—1650	118
$Ca_3(PO_4)_2$	$CaO-P_2O_5-ZnO$	1048—1730	93
$Ca_3(PO_4)_2$	$CaO-P_2O_5-MgO$	1175—1730	1
$Ca_4(PO_4)_2O$	$CaO-P_2O_5-Al_2O_3$	1570—1635	155
$Ca_3(PO_4)_2$	$CaO-P_2O_5-Al_2O_3$	1275—1730	155
$Ca_2P_2O_7$	$CaO-P_2O_5-Al_2O_3$	970—1353	155
$Ca_3(PO_4)_2$	$CaO-P_2O_5-SiO_2$	1555—1730	157
$Ca_4(PO_4)_2O$	$CaO-P_2O_5-SiO_2$	1570—1670	9
$Ca_3(PO_4)_2$	$CaO-P_2O_5-MgO-SiO_2$	1246—1730	143
$Ca_3(PO_4)_2$	$CaO-P_2O_5-Al_2O_3-SiO_2$	1280—1730	156
(Na,CO_3) apatite	$CaO-P_2O_5-H_2O-Na_2O-CO_2$	780—?	147

Table 3
SOLID SOLUTIONS OF "FOREIGN COMPOUNDS" INTO A SOLID CALCIUM PHOSPHATE (Ca/P ≥ 1) AT HIGH TEMPERATURES IN DIFFERENT SYSTEMS

Calcium phosphate	"Foreign compound"	Maximum content wt. %	Ref.
β-$Ca_2P_2O_7$	Extra P_2O_5	7	78
β-$Ca_3(PO_4)_2$	Extra P_2O_5	5	187
α-$Ca_3(PO_4)_2$	Extra P_2O_5	10	187
β-$Ca_3(PO_4)_2$	$CaNaPO_4$	20	2
β-$Ca_3(PO_4)_2$	$Zn_3(PO_4)_2$	18	93
β-$Ca_3(PO_4)_2$	β-$Sr_3(PO_4)_2$	100	144
β-$Ca_2P_2O_7$	$Mg_2P_2O_7$	7	107
β-$Ca_3(PO_4)_2$	$Mg_3(PO_4)_2$	10	1
β-$Ca_3(PO_4)_2$	Ca_2SiO_4	1	55
α-$Ca_3(PO_4)_2$	Ca_2SiO_4	4	55
$\bar{\alpha}$-$Ca_3(PO_4)_2$	Ca_2SiO_4	100	55
$Ca_{10}(PO_4)_6(OH)_2$	$Ca_{10}(PO_4)_6F_2$	100	145
	$Ca_{10}(PO_4)_6CO_3$	100	14
$Ca_{10}(PO_4)_6(CO_3)$	$CaCO_3$	8	147
	$Ca_{8.5}Na_{1.5}(PO_4)_{4.5}(CO_3)_{2.5}$	100	147
$Ca_{10}(PO_4)_6(OH)_2$	$Sr_{10}(PO_4)_6(OH)_2$	100	77
	$Pb_{10}(PO_4)_6(OH)_2$	60(?)	182

$CaO-P_2O_5$;ZnO at 1029°C. Thus these systems are the most suitable for carrying out the process of densification by liquid sintering.

As far as $Ca_3(PO_4)_2$ is concerned, the $\bar{\alpha}$-$Ca_3(PO_4)_2$ structure is stabilized by incorporation of Ca_2SiO_4, whereas the β-$Ca_3(PO_4)_2$ phase is stabilized by incorporation of ZnO, MgO, or SrO_2. Therefore, the question is left open as to whether this latter phase is really β-$Ca_3(PO_4)_2$ or whitlockite.

More complex systems such as $CaO-P_2O_5-H_2O-CO_2-MgO-Na_2O$ may result in situations where a magnesium-containing whitlockite and a Na- and CO_3-containing apatite are both in equilibrium with one liquid phase at a relatively low temperature. As will be shown in Section IV, such a mixture of solid phases resembles the tooth and bone mineral very well. Therefore, these combinations should be tested in order to obtain dense calcium phosphate bodies resembling the calcified tissues, simply by sintering at relatively low temperatures in CO_2, and to apply them for implantation purposes.

B. Stability of Calcium Phosphates in Contact With Aqueous Solutions

Solid calcium phosphates in contact with an aqueous solution can exchange ions with that aqueous solution on their surface. The equilibrium that may ultimately result is restricted between the aqueous phase and the surface of the solid phase. The bulk of the solid phase is not involved as solid-state diffusion is excluded at those ordinary temperatures. These are low in comparison with the melting point of the solids.

It should also be mentioned that the ions which can be exchanged at the surface of a solid calcium phosphate are involved in several secondary reactions occurring in the aqueous phase. The dissociation constants of some relevant ionic complexes are given in Table 4.

Another remark that should be made is that some calcium phosphates are not stable when in contact with certain aqueous solutions. Initial dissolution of the solid phase may then result in the re-precipitation of another solid calcium phosphate phase. The latter may form as separate particles so that the transformation of the first solid into the second can go to completion. It may also form as a very thin impervious layer on the particles of the first solid so that the partial transition into the second solid may be barely detectable.

As far as calcium phosphates for implantation purposes are concerned, it should be kept in mind that the body fluids which come in contact with them are refreshed continuously by the body's physiology. This means that "unbiological" additions incorporated in the synthetic calcium phosphates for the sake of easy technical production, tend to vanish from the surrounding body fluid first as they are removed preferentially through metabolic activity.

1. Solid Phase—Aqueous Phase Equilibria in the System $CaO-P_2O_5-H_2O$

The solubility product of $CaHPO_4$ was determined most accurately by McDowell, Brown, and Sutter,[109] and that of $CaHPO_4 \cdot 2H_2O$ by Gregory, Moreno, and Brown.[66] The latter is not affected by the presence of Na^+ and Cl^- ions.[130] The solubility product of whitlockite has not been determined as yet, although an estimate is derived[35] from the work of Meyer and Eanes.[110] Similarly, an estimate of the solubility product of octocalcium phosphate was obtained.[114]

Several authors have investigated the solubility product of hydroxyapatite. Depending on the method of preparation of the samples, the negative logarithm of its solubility product can vary from about 98 to 118.[115,119,137] This variation is about 10 times as large as can be accounted for by surface effects due to different particle sizes.[87] The best constant values for the solubility product of hydroxyapatite were obtained by equilibration of refluxed or carefully washed samples.[6,26,27,178,189] The best value appears to be that of Verbeeck[178]: $pK_1 = 117.2$. Chuong[26] found that the value does not depend on the presence of Cl^- ions.

The calcium content of the aqueous solution in these equilibria is a measure for the solubility of these calcium phosphates in the ternary system. Therefore, Figure 2, in which this calcium content is given as a function of pH, is called a solubility diagram. It is observed from this figure that the solubility of hydroxyapatite increases by a factor between 6 and 10 for every decrease of pH by 1 unit. This is mainly due to the fact

Table 4
NEGATIVE LOGARITHMS OF THE DISSOCIATION CONSTANTS OF CERTAIN IONIC COMPLEXES RELEVANT TO AQUEOUS SOLUTIONS IN CONTACT WITH A SOLID CALCIUM PHOSPHATE

Species	pK_1	pK_2	pK_3	Ref.
H_3PO_4	2.12	7.21	12.35	106
$CaH_2PO_4^+$	0.71			66
$CaHPO_4$	2.41			66
$CaPO_4^-$	6.46			69
$MgHPO_4$	2.5			69
$NaHPO_4^-$	1.15			69
$KHPO_4^-$	1.04			69
H_2CO_3	6.37	10.25		61
$CaHCO_3$	3.26			61
$CaCO_3$	3.2			61
$MgHCO_3^-$	1.16			61
$MgCO_3$	3.4			61
$NaHCO_3$	−0.25			61
$NaCO_3^-$	1.27			61

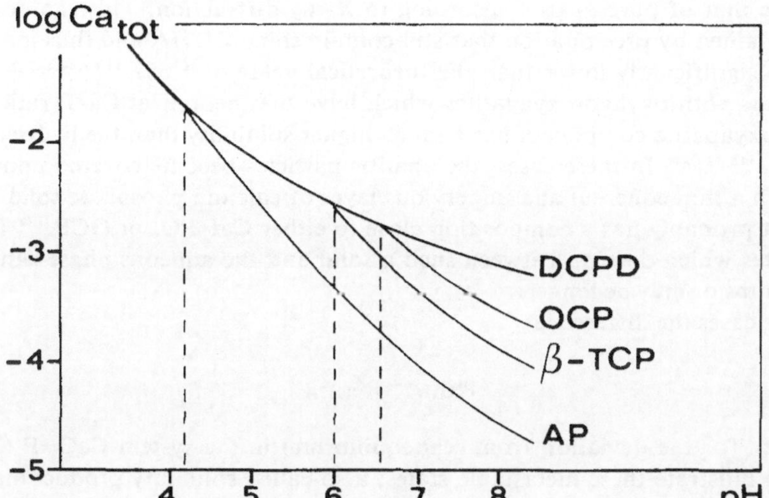

FIGURE 2. Calcium content of aqueous solutions of calcium phosphates in diluted phosphoric acids as a function of pH. (From Driessens, F. C. M., *Bull. Soc. Chim. Belg.*, 89, 663, 1980. With permission.)

that phosphoric acid is a weak acid. From this same figure, it is observed that OHA is the stable phase in the range pH > 4.2.

If apatite samples are not prepared in the proper way, the values found after their equilibration in an aqueous solution for the negative logarithm of their ionic product given by

$$pI_{OHA} = 10\,pCa + 6\,pPO_4 + 2\,pOH$$

can deviate much from 117.2 as mentioned before, although the structure of the solid

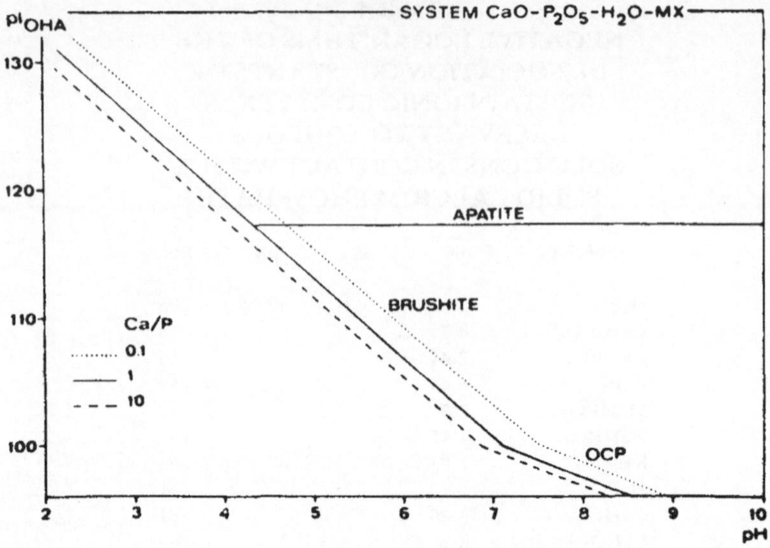

FIGURE 3. Solubility product diagram of aqueous solutions of calcium phosphates. (From Driessens, F. C. M., van Dijk, J. W. E., and Borggreven, J. M. P. M., *Z. Naturforsh.*, 34c, 165, 1979. With permission.

seems to be that of pure apatite according to X-ray diffraction. This can occur with samples obtained by precipitation that still contain some DOHA and thus may have a Ca/P ratio significantly lower than the theoretical value of 1.667.[12,83,84,137] However, it also occurs with oxyhydroxyapatites which have the theoretical Ca/P ratio, but in which the oxyapatite component has a much higher solubility than the hydroxyapatite component.[115,119,176] In these cases, the apatitic particles become covered upon equilibration with a thin coherent and impervious layer of calcium phosphate solid solution which most probably has a composition close to either $CaHPO_4$ or OCP.[180] The metastable states which develop between such a solid and the aqueous phase can last for several months or may be longer.

For these cases the difference

$$pI_{OHA} - pK_{OHA}$$

is a measure for the deviation from real equilibrium in the system $CaO-P_2O_5-H_2O$. In order to illustrate these metastable states, a so-called solubility product diagram is most illustrative. This is given in Figure 3 by plotting pI_{OHA} vs pH. The curves in this figure are not exactly straight, except for that of pure hydroxyapatite. MX in this figure represents an indifferent electrolyte consisting of ions M^+ and X^- not affecting the solid phase and added each in amounts to make the system electroneutral. The curve for whitlockite is below the indicated field of metastable apatite and runs through about pI_{OHA} 91 at pH 7 and pI_{OHA} 96 at pH 5, which is below the field of "relative supersaturation". However, the curve for DOHA runs through pI_{OHA} 101.0 at pH 7 and pI_{OHA} 103.5 at pH 5, which is partially within the field of "relative supersaturation". The curve for $CaHPO_4$ is not included as this phase is rarely formed in aqueous systems. It runs very close to that of brushite $CaHPO_4 \cdot 2H_2O$.

2. Solid Phase—Aqueous Phase Equilibria in the Systems $CaO-P_2O_5-H_2O-CO_2$ and $CaO-P_2O_5-H_2O-CO_2-Na_2O$

Clark and Turner[28] studied the equilibrium reached in aqueous solutions after the

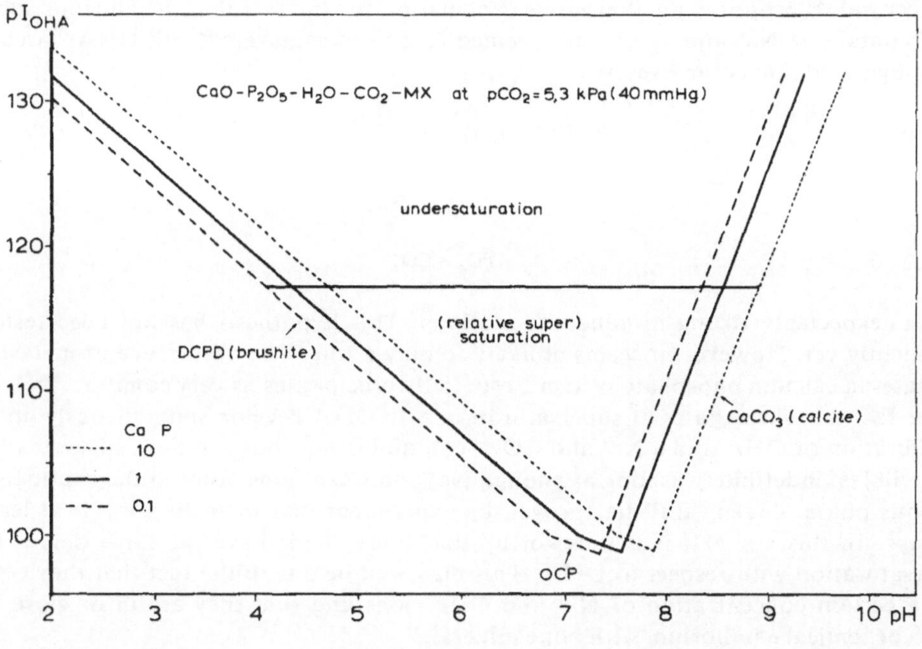

FIGURE 4. Solubility product diagram for the system containing (Na and) CO_3. (From Driessens, F. C. M., van Dijk, J. W. E., and Borggreven, J. M. P. M., *Z. Naturforsh.*, 34c, 165, 1979. With permission.)

reaction of H_3PO_4 with $CaCO_3$ in the form of calcite during a period of up to two weeks under certain partial CO_2 pressures which fixed the pH at a value between 5 and 9. They found that the aqueous solutions were saturated with brushite DCPD in that range. It is not clear whether the supersaturation with OHA is due to the incorporation of CO_3^{2-} ions into an OCP-like or DOHA-like calcium phosphate. However, the obtained solubilities apparently are located in the field of "relative supersaturation" of Figures 3 and 4 which obtain a new borderline at the side of high pH due to the fact that $CaCO_3$ becomes a stable solid phase.

Literature data on equilibrations between solid calcium phosphates having the apatite structure and aqueous solutions containing Na^+ and CO_3^{2-} ions have been presented by several authors.[7,65,68,79,88,104,149] These data were reexamined by Driessens, van Dijk, and Borggreven.[34] They calculated the negative logarithm of the ionic products for OHA, DCPD, DOHA, and OCP. The calculations showed that none of the equilibrated solutions was supersaturated or saturated with calcite. However, all solutions were supersaturated with OHA. This proves the validity of Figure 4 for this system if one takes into account the consideration that the aqueous solutions which appeared to be slightly supersaturated with calcite might have been so due to the fact that the reprecipitated calcite phase in such systems is very finely dispersed.

As shown by Ericsson,[50] a certain sample of OHA might well equilibrate according to the solubility product principle in a H_3PO_4-containing aqueous solution. Yet, this same OHA sample will form an aqueous solution supersaturated with OHA when the original aqueous solution contains Na^+ and CO_3^{2-} ions. This might happen by the extraordinarily strong chemisorption of primarily CO_3^{2-} ions.[121] This probably results in a coherent, impervious layer of carbonated or Na- and CO_3-containing apatite onto the OHA particles. Apparently, such an apatite layer may be conceived rather as a sort of DOHA, because above about pH 6.5, its solubility behavior is very close to that of DOHA.[34]

Greenwald[65] proposed an alternative explanation for the fact that the aqueous solutions containing Na^+ and CO_3^{2-} ions seemed to be supersaturated with OHA. Greenwald suggested that complexes like

$$Ca_2 \cdot HPO_4 \cdot CO_3$$

and

$$Ca_2 \cdot PO_4 \cdot CO_3^-$$

were unexpectedly strong in aqueous solutions. This hypothesis has not been tested sufficiently yet. However, it seems unlikely to play a role. The occurrence of metastable states in calcium phosphate systems, especially precipitates, is very common.[180]

The fact that these states of supersaturation with OHA develop spontaneously upon equilibration of OHA and a Na^+ and CO_3^{2-}-containing aqueous solution, suggests that they will last indefinitely as long as enough Na^+ and CO_3^{2-} ions occur in the contingent aqueous phase. Greenwald[65] has proven by experiment that their duration is at least 60 days. In this context, it is noteworthy that body fluids have the same degree of supersaturation with respect to OHA. This may well be due to the fact that they contain a certain concentration of Na^+ and CO_3^{2-} ions and that they are in or close to physicochemical equilibrium with bone mineral.[35]

According to the analysis of literature data about equilibrium in the system $CaO-P_2O_5-H_2O-Na_2O-CO_2-MX$ by Driessens, van Dijk, and Borggreven,[34] the data below about pH 6.5 suggest that a surface layer of not a carbonated apatite, but one similar to DCPD or OCP[180] covers the underlying apatite and controls the solubility behavior in that pH range. Bone mineral and also tooth enamel mineral display the same behavior, when equilibrated in slightly acidic aqueous solutions.[38]

3. *Solid Phase—Aqueous Phase Equilibria in the System $CaO-P_2O_5-H_2O-MgO$*

As mentioned in Section II.A., Mg^{2+} ions are incorporated only in the whitlockite phase, not in the other calcium phosphates relevant to aqueous systems. Equilibrations of calcium phosphates in Mg^{2+}-containing aqueous solutions were carried out sparingly and most were in older studies especially close to physiological conditions of Mg^{2+} concentration and pH.[57,79,95,96] Ionic products for the different calcium phosphates were calculated for these literature data.[35] Some aqueous solutions were supersaturated with OCP and even more with DOHA. Thus, these combined data are evidence[33] that the Mg^{2+} ions do not interact with OCP and DOHA. The solubility of whitlockite is probably decreased drastically the more Mg^{2+} ions are incorporated. Precipitates containing little magnesium had a pI_{OHA} value between 90 and 93 close to saturation with ACP (which probably is a fine disperse Ca-whitlockite), whereas those containing magnesium in amounts close to the composition $Ca_9Mg(HPO_4)(PO_4)_6$ show pI_{OHA} values between 98 and 102 (Kuyper,[96] and Table 1 within). Therefore, magnesium incorporation up to the level of Equation 32 brings the degree of supersaturation of aqueous solutions in contact with whitlockite probably down to a level equalling that of carbonated apatite, Na- and CO_3-containing apatite, and DOHA at the physiological pH of 7.4.

4. *Solid Phase—Aqueous Phase Equilibria in the System $CaO-P_2O_5-H_2O-HF$*

As mentioned in Section II.A., apatite is the only calcium phosphate in which fluoride can be incorporated, whereas care must be taken to avoid the formation of CaF_2 as a second solid phase. Recent determinations of the solubility product of CaF_2 resulted in mutually very close results. According to Macaskill and Bates,[105] the negative

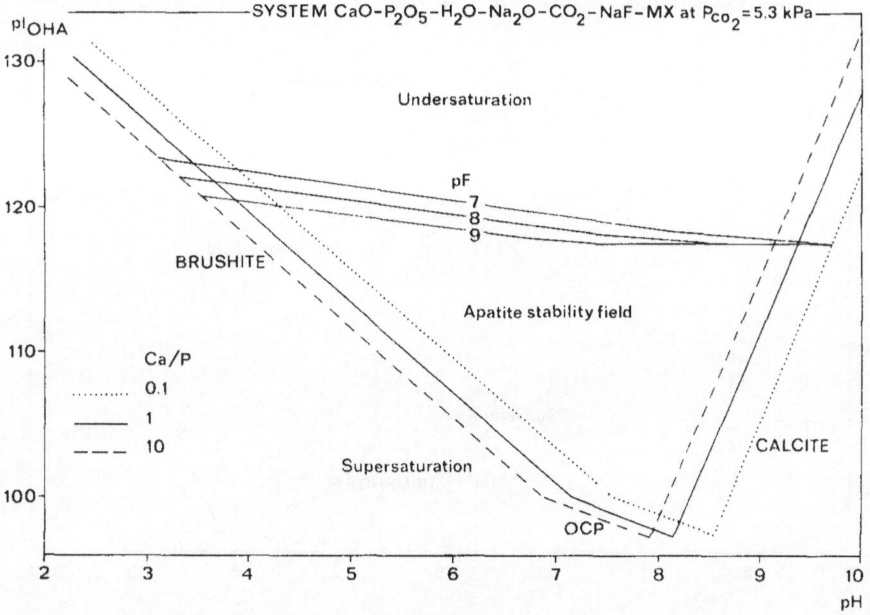

FIGURE 5. Solubility product diagram of aqueous solutions of calcium phosphates in the presence of low concentrations of fluoride. (From Driessens, F. C. M., van Dijk, J. W. E., Borggreven, J. M. P. M., and Verbeeck, R. M. H., *J. Biol. Buccoale*, 8, 239, 1980. With permission.)

logarithm is 10.50, whereas Verbeeck, Khan, and Thun[175] as well as McCann[106] found 10.45.

The solubility product of pure fluorapatite (FA) has been determined by several authors. Hagen[71] found values for its negative logarithm between 116.8 and 118.8 on a mineral sample. McCann[106] found values between 117.2 and 121.2 for precipitated samples. Moreno, Kresak, and Zahradnik[117] found 121.0 ± 0.15 for a sample prepared by very careful precipitation. Farr and Elmore[63] found 120.86 ± 0.15 for a similar sample. Wier, Chien, and Black[190] found values ranging from 117.7 to 120.5 for an FA coating on OHA particles, whereas Verbeeck, Thun, and Driessens[176] found a value of 121.7 ± 0.5 for an FA coating on fluoroxyhydroxyapatite particles.

For these reasons, the presence of fluoride ions in the aqueous phase can affect the form of the solubility product diagram of calcium phosphates due to the formation of either FA or CaF_2 or both.[97] As shown in Figures 5 and 6 this form depends mainly on the level of pF in the aqueous solution. At high pF the field of "relative supersaturation" is enlarged due to FA formation, at lower pF, this field shrinks due to CaF_2 formation. The validity of these diagrams has been proven by equilibrations of fluoroxyhydroxyapatites.[177]

5. Solid Phase—Aqueous Phase Equilibria in the System $CaO-P_2O_5-H_2O-SrO$

As shown in Section II.B., strontium can be incorporated in both the apatite and the whitlockite structure. The solubility product of pure SrOHA has been determined recently.[174,183] These investigations enabled the construction of solubility product diagrams for the system $SrO-P_2O_5-H_2O$ in which the phases $SrHPO_4$, $Sr_6H_3(PO_4)_5 \cdot 3H_2O$, and SrOHA play a role very similar to $CaHPO_4 \cdot 2H_2O$, OCP, and OHA, respectively in the system $CaO-P_2O_5-H_2O$.[181]

These data predict only that little strontium is incorporated into apatite by precipitation from an aqueous solution containing comparable amounts of Ca^{2+} and Sr^{2+} ions. There is a strong preference for Ca^{2+} ions to be taken up in the precipitate.

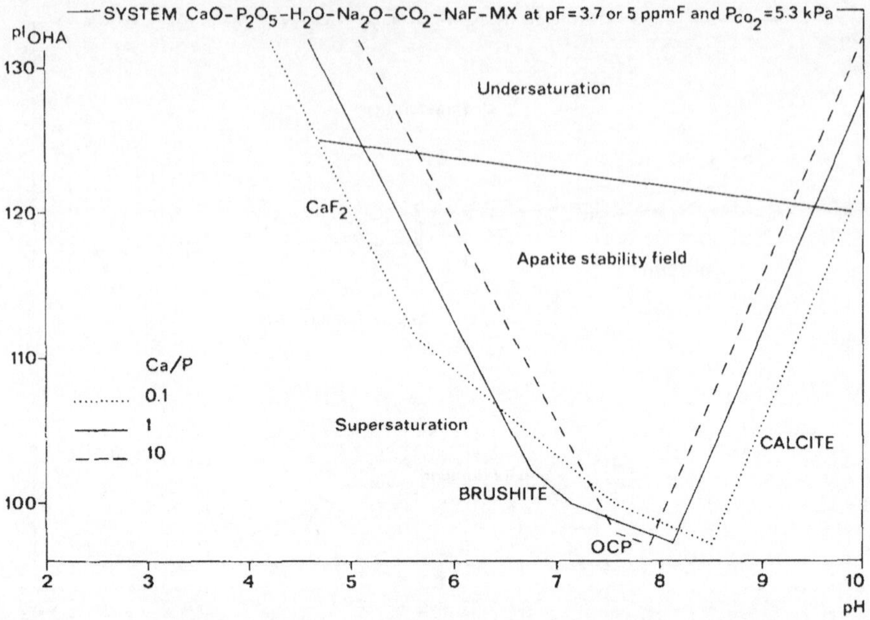

FIGURE 6. Solubility product diagram of aqueous solutions of calcium phosphates in the presence of a higher concentration of fluoride. (From Driessens, F. C. M., van Dijk, J. W. E., Borggreven, J. M. P. M., and Verbeeck, R. M. H., *J. Biol. Buccale*, 8, 239, 1980. With permission.)

Analysis of the biological effects of strontium on tooth tissues and bone suggests that under in vivo conditions, strontium may be incorporated in a whitlockite phase.[40,41] This interpretation is supported by the fact that strontium incorporation stabilizes the whitlockite structure.[144]

6. Additional Remarks on Solid Phase—Aqueous Phase Equilibria

A few equilibrations have been reported on calcium phosphates containing other additives. Chiranjeevi Rao[25] claimed that the phosphate solubility of Ca-Zn hydroxylapatites in diluted phosphoric acid buffers in the range pH 5 to pH 8 decreases with increasing zinc content up to 60 mol percent zinc hydroxylapatite. This investigation is somewhat doubtful as the author claims to have prepared pure zinc hydroxylapatite, which has never been obtained by other investigators. As shown in the foregoing section, whitlockite or β-TCP is the calcium phosphate which incorporates some zinc preferentially.

In a similar study, Chiranjeevi Rao[23] showed that the presence of Fe^{2+} in an aqueous solution of calcium hydroxylapatite decreased the phosphate solubility. Chiranjeevi Rao interpreted this as being due to the formation of Fe-Ca hydroxylapatites. It seems more likely that precipitates of iron phosphates have been formed.

The situation is somewhat different for the system of CaPb hydroxylapatites.[24] There, homogeneous solid solutions seem to have been prepared first, and were characterized by X-ray diffraction. Their phosphate solubility decreased drastically up to about 60% lead hydroxylapatite.

Similarly, Nayak and Chiranjeevi Rao[120] determined the phosphate solubility of Ca-Cd hydroxylapatites and found a decrease up to 40 mol percent cadmium hydroxylapatite. In such studies much attention should be paid to the eventual precipitation of phases other than calcium phosphates, as the extensive studies on the system $CaO-P_2O_5-H_2O-SrO$ have shown.[176] Another aspect is that many of these apatites were prepared by ceramic techniques. As shown in the previous section, they are then transformed into oxyhydroxyapatites.[150,179] The oxyapatite component makes these apatites more unstable in contact with aqueous solutions than any of the calcium phosphates mentioned in Figures 2 through 4.

The situation is the reverse for the solubility behavior of the high temperature β-TCP in aqueous solutions in comparison with that of the low temperature whitlockite form. Gregory, Moreno, Patel, and Brown[67] found that β-TCP was less soluble than brushite and octocalcium phosphate in pH range 6.0 to 7.5, whereas in Section III.B.1., it was concluded that whitlockite is more soluble than brushite and octocalcium phosphate in that range. Thus, a high temperature treatment of whitlockite will increase its stability towards aqueous solutions.

As soon as the concentration of F^- ions in an aqueous solution is larger then about 10^{-7} mol l^{-1}, fluorapatite is more stable than hydroxyapatite in contact with that aqueous solution.[37] However, even at high Cl^- concentrations, chlorapatite remains less stable than hydroxyapatite. This can be derived from thermodynamical data[64,112] and is consistent with the fact that Cl^- is not incorporated in apatite precipitated from aqueous solutions containing this ion.

Rhenanite phases are a major component of some phosphate fertilizers because of their relatively high solubility. They are, therefore, very unstable for aqueous solutions.

A semiquantitative study of solubility and transformation of calcium phosphates during contact with aqueous solutions was carried out by Schleede, Schmidt, and Kindt[148] on brushite, β-TCP, tetracalcium phosphate, silicocarnotite, and a number of phosphate fertilizers. Silicocarnotite appeared to be the most stable calcium phosphate.

Finally, it may be mentioned that this variety of solubilities of the different calcium phosphates enables the selection of certain calcium phosphates for specific implantation purposes depending on the rate of resorption that is desirable for that specific application.

IV. PHASE COMPOSITION OF THE MINERAL IN CALCIFIED TISSUES

Apart from the main constituents Ca and PO_4, the minerals of bone, dentin, and tooth enamel contain considerable amounts of CO_3, Na, Mg, and smaller amounts of K, Cl, and F. Typical compositions are given in Table 5.[35] The minerals of bone and dentin also contain a considerable amount of citrate.

In the past various concepts have been advanced about the phase composition of these minerals. Since de Jong[82] proved that the X-ray diffraction pattern of bone was similar to that of apatite, many authors thought that these biominerals were single-phase apatites derived either from OHA[89] or from carbonato-apatite $Ca_{10}(PO_4)_6CO_3$.[138]

Dallemagne and Melon[29] discovered that the chemical properties of the mineral of bone and dentin and of about 40% of the mineral of enamel were similar to those of DOHA. On that basis they proposed DOHA as the prototype of that mineral, still thinking of it as a single-phase apatite containing all the components mentioned in Table 1. Posner and Stephenson[131] were critical about this view for the whole mineral of tooth enamel.

More recently, other concepts and additional experimental data have been put forward, especially for the mineral in bone. Brown[21] proposed that the apatitic crystals could be covered with or even be interlayered by OCP. Evidence that an amorphous phase is a major component of bone mineral came from electron microscopic studies[54,135] and from X-ray diffraction studies.[72,133] In young bone samples of 5-day

Table 5
AVERAGE COMPOSITION OF THE MINERAL IN SPECIMENS OF HUMAN CALCIFIED TISSUES AS FAR AS THE MAJOR COMPONENTS ARE CONCERNED

Component	Bone	Dentin	Enamel
Ca	36.7	36.8	36.4
P	16.0	18.0	17.1
CO_3	8.0	6.55	3.40
Na	0.77	0.38	0.64
Mg	0.46	1.24	0.43
F	0.04	0.03	0.01
Ca/P	1.77	1.58	1.64

From Driessens, F. C. M., *Bull. Soc. Chim. Belg.*, 89, 663, 1980. With permission.

old rats, the percentage amorphous mineral was 69%, which decreased to 36% in the same bones of 70-day old rats.[160] The particles of the amorphous phase are still intact after hydrazine de-proteination of bone and can be observed directly by electron microscopy.[164] Investigations on single osteons[5] have shown that apart from crystalline platelets,[81,164] needle-like crystals also occur in bone. This suggests that in addition to the amorphous phase, there may be at least two different crystalline phases in bone mineral.

Driessens[35] and Driessens and Verbeeck[42] have analyzed the available literature data about the composition of samples of the mineral in bone, dentin, and tooth enamel. Their analysis was carried out on the basis of the general phenomenon that the main impurity ions mentioned in Table 4 (according to the data presented in Sections II and III) have a strong preference for incorporation in different calcium phosphates. Therefore, it follows from Section II.A. that Mg^{2+} can be incorporated in whitlockite which then preferably has the formula

$$Ca_9\,Mg\,(HPO_4)\,(PO_4)_6 \qquad (32)$$

The most probable way in which Na and CO_3 occur in the mineral is in the form of the Na- and CO_3-containing apatite preferably having the composition

$$Ca_{8.5}\,Na_{1.5}\,(PO_4)_{4.5}\,(CO_3)_{2.5} \qquad (33)$$

whereas citrate is supposed to have a preference for an OCP phase probably having the composition[36]

$$Ca_{7.5}\,(Citr.)\,(PO_4)_4 \cdot 5\,H_2O \qquad (34)$$

These phases are supposed to occur in the minerals of both bone and dentin and tooth enamel.

A. The Rest Phase in Tooth Enamel Mineral

Driessens and Verbeeck[42] have proposed that the Na- and CO_3-containing apatite accounts for the different sodium contents of samples of tooth enamel and that Mg-whitlockite accounts for the different magnesium contents of the same samples. From the various analytical data for the content of Ca, PO_4, Na, and Mg of enamel samples found in the literature, they calculated the Ca/P ratio of an eventual "rest phase" They found 1.662 ± 0.028 which equals that of OHA.

They also found that Equation 33 does not completely account for the CO_3 content of these samples of tooth enamel mineral. They suppose that the "excess" carbonate is contained in the OHA phase. By comparison of the Na, CO_3, Mg, Cl, and F gradients reported to occur in the tooth enamel of single teeth they concluded that this OHA phase also contains the Cl content of the mineral, which gives it about the composition

$$Ca_{10}\ (PO_4)_6\ (OH)_{1.59}\ (CO_3)_{0.15}\ (Cl)_{0.1}\ (F)_{0.01} \tag{35}$$

Such compositions have been reported for synthetic OHA preparations made without the exclusion of carbon dioxide.[49,52,166] This slightly carbonated hydroxyapatite phase is the one which incorporates F^- ions if tooth enamel is exposed to fluoride-containing aqueous solutions. The relative amounts of Mg-whitlockite, of Na- and CO_3-containing apatite and of slightly carbonated hydroxyapatite in human tooth enamel mineral vary slightly around 10%, 20%, and 70%, respectively.

B. The Rest Phase in the Mineral of Bone and Dentin

Similarly, Driessens[35,36] assumed that Na- and CO_3-containing apatite accounts for the sodium content of the different samples of the mineral of bone and dentin mentioned in the literature, Mg-whitlockite for the magnesium content and citrate containing OCP for the citrate content. In these considerations, 11 dentin samples and over 100 bone samples were involved. From the Ca, PO_4, Na, and Mg content of these samples, the Ca/P ratio of an eventual "rest phase" were calculated. For dentin, a Ca/P ratio of 1.969 ± 0.062 and for bone, a Ca/P ratio of 1.970 ± 0.022 was found for this "rest phase".

In both cases a considerable part of the CO_3 content was not accounted for by the Na- and CO_3-containing apatite. It was supposed that the "excess" carbonate was contained in the "rest phase", which in both cases became the compound

$$Ca_9\ (PO_4)_{4.5}\ (CO_3)_{1.5}\ (OH)_{1.5} \tag{36}$$

This has been prepared synthetically by precipitation from a carbonate-containing aqueous solution.[97] It appears to be apatitic as well. This phase is supposed to incorporate the F^- ions which can be taken up by bone and dentin mineral under in vivo conditions.[36]

The relative amounts of citrate-containing OCP, of Na- and CO_3-containing apatite, of Mg-whitlockite, and of the heavily carbonated hydroxyapatite of Equation 36 in the mineral of dentin vary slightly around 3%, 20%, 47%, and 30%, respectively, and in the mineral of bone around 10%, 13%, 18%, and 59%, respectively.

V. PHYSICOCHEMICAL FACTORS RELATED TO BODY FLUIDS AND BIOMINERALS

Body fluids such as saliva and plasma are supersaturated with OHA, whereas according to the best estimations, they are saturated simultaneously with Na- and CO_3-containing apatite and Mg-whitlockite,[33,35] and probably to citrate-containing OCP and heavily carbonated hydroxyapatite as well. This supports the hypothesis that calcium homeostasis is regulated by the physicochemical equilibrium between bone mineral and blood plasma.[35] Similarly, evidence was obtained recently for the fact that strontium homeostasis is reached when using strontium-containing diets such that it is controlled by the same physicochemical equilibrium between bone mineral and blood plasma.[4] These considerations certainly apply on the systemic level. It should be realized that in this physicochemical equilibrium at least the ions of Ca^{2+}, PO, Na^{+}, Mg^{2+}, carbonate, and citrate and perhaps those of Cl^{-}, SO_4^{2-}, Sr^{2+}, and F^{-} are involved. This means that their metabolism is controlled by the composition of bone mineral.

On the cellular level, osteocytes can exert either osteoblastic or osteoclastic activity leading, among others, to a continuous turnover of bone and its mineral. As far as the physicochemistry of the phases in bone mineral is considered, the processes of local dissolution and local reprecipitation of bone mineral can be comprehended only by assuming that the cellular activity of the osteocytes leads to a local increase or a local decrease of the pH in the bone fluid around osteoblasts and osteoclasts, respectively. These local pH gradients will not leave unaffected any synthetic calcium phosphate implant. However, the rate of its breakdown and its eventual replacement by bone may depend very much on the type of additives used for its fabrication and on its pure structure. Much work has still to be done to develop the optimum composition for each field in which the use of calcium phosphate implants may be indicated clinically.

ACKNOWLEDGMENT

The author is indebted to Professor Rieck of the Technological University of Eindhoven and to Professor F. Verbeek of the State University of Ghent for their support of and stimulating interest in the study of calcium phosphates. This work could not have been written without the essential and valuable contributions of Dr. R. M. H. Verbeeck of the State University of Ghent to many of the projects carried out by the author in the field of calcium phosphates during the past few years.

REFERENCES

1. Ando, J., Phase diagrams of $Ca_3(PO_4)_2-Mg_3(PO_4)_2$ and $Ca_3(PO_4)_2-CaNaPO_4$ systems, *Bull. Chem. Soc. Jpn.*, 31, 201, 1958.
2. Ando, J. and Matsuno, S., $Ca_3(PO_4)_2-CaNaPO_4$ system, *Bull. Chem. Soc. Jpn.*, 41, 342, 1968.
3. Arends, J., Schuthof, J., van der Linden, W. H., Bennema, P., and van den Berg, P. J., Preparation of pure hydroxyapatite single crystals by hydrothermal recrystallization, *J. Cryst. Growth*, 46, 213, 1979.
4. Arnold, P. W., The nature of precipitated calcium phosphates, *Trans. Faraday Soc.*, 46, 1061, 1950.
5. Ascenzi, A., Bonuci, E., Ripamonti, A., and Roveri, N., X-ray diffraction and electron microscope study of osteons during calcification, *Calcif. Tissue Res.*, 25, 133, 1978.
6. Avnimelech, Y., Moreno, E. C., and Brown, W. E., Solubility and surface properties of finely divided hydroxyapatite, *J. Res. Natl. Bur. Stand.*, 77A, 149, 1973.
7. Bachra, B. N., Trautz, O. R., and Simon, S. L., A precipitation diagram for the system calcium-carbonate-phosphate and the heterogeneous nucleation of solids in the metastability region, in *Advances in Fluoride Research and Dental Caries Prevention*, Vol. 3, Hardwick, J. L., Held, H. R. and König, K. G., Eds., Pergamon Press, Oxford, 1965, 101.
8. Bachra, B. N., Trautz, O. R., and Simon, S. L., The effect of magnesium and fluoride ions on the spontaneous precipitation of calcium carbonates and phosphates, *Arch. Oral Biol.*, 10, 731, 1965b.
9. Barrett, R. L. and McCaughey, E. J., The system $CaO-SiO_2-P_2O_5$, *Am. Mineral.*, 27, 687, 1942.
10. Bassett, H., The phosphates of calcium. IV. The basic phosphates, *J. Chem. Soc.*, 111, 620, 1917.
11. Bassett, H., The phosphates of calcium. V. Revision of the earlier space diagram, *J. Chem. Soc.*, 2949, 1958.
12. Bell, L. G., Mika, H., and Kruger, B. J., Synthetic hydroxyapatite solubility product and stoichiometry of dissolution, *Arch. Oral Biol.*, 23, 329, 1978.
13. Billhardt, H. W., Phasenbildungen im System Soda-Kalk, *Glastech. Ber.*, 42, 272, 1969.

14. Bonel, G., Contribution à l'étude de la carbonatation des apatites, *Ann. Chim. (Paris)*, 7, 65, 1972.
15. Bonel, G., Labarthe, J. C., and Vignoles, C., Contribution à l'étude structurale des apatites carbonatées de type B, *Colloq. Int. CNRS*, 230, 117, 1975.
16. Boskey, A. L. and Posner, A. S., Formation of hydroxyapatite at low supersaturation, *J. Phys. Chem.*, 80, 40, 1976.
17. Boulet, M., Marier, J. R., and Rose, D., Effect of magnesium on formation of calcium phosphate precipitates, *Arch. Biochem. Biophys.*, 96, 629, 1962.
18. Bredig, M. A., Franck, H. H., and Füldner, H., Substanzen aus Calciumoxyd und Phosphorsäure, *Z. Elektrochem.*, 39, 959, 1933.
19. Brown, W. E., Crystal structure of octocalcium phosphate, *Nature (London)*, 196, 48, 1962.
20. Brown, W. E., Smith, J. P., Lehr, J. R., and Frazier, A. W., Crystallographic and chemical relations between octacalcium phosphate and hydroxyapatite, *Nature (London)*, 136, 1050, 1962.
21. Brown, W. E., Crystal growth of bone mineral, *Clin. Orthop.*, 44, 205, 1966.
22. Cameron, F. K. and Hurst, L. A., The action of water and saline solutions upon certain slightly soluble phosphates, *J. Am. Chem. Soc.*, 26, 885, 1904.
23. Chiranjeevi Rao, S. V., Some aspects of uptake of iron by calcium hydroxyl apatite and effect of such incorporation on the solubility of bone mineral, *J. Inst. Chem. Calcutta*, 47, 17, 1975.
24. Chiranjeevi Rao, S. V., Physico-chemical studies of calcium-lead hydroxyl apatites, *J. Indian Chem. Soc.*, 53, 352, 1976.
25. Chiranjeevi Rao, S. V., pH dependence of solubility of calcium-zinc hydroxylapatites, *J. Indian Chem. Soc.*, 53, 585, 1976b.
26. Chuong, R., Experimental study of surface and lattice effects on the solubility of hydroxyapatite, *J. Dent. Res.*, 52, 911, 1973.
27. Clark, J. S., Solubility criteria for the existence of hydroxylapatite, *Can. J. Chem.*, 33, 1696, 1955.
28. Clark, J. S. and Turner, R. C., Reactions between solid calcium carbonate and orthophosphate solutions, *Can. J. Chem.*, 33, 665, 1955.
29. Dallemagne, M. J. and Melon, J., Formation of brushite from the salts of bones and teeth, *Bull. Soc. Chim. Biol.*, 27, 597, 1945.
30. Denissen, H. W., van Dijk, H. J. A., de Groot, K., Klopper, P. J., Vermeiden, J. P. W., and Gehring, A. P., Biological and mechanical evaluation of dense calcium hydroxy-apatite made by continuous hot pressing, in *Mechanical Properties of Biomaterials*, Hastings, G. W. and Williams, D. F., Eds., John Wiley & Sons, New York, 1980, 489.
31. Dickens, B. and Brown, W. E., The crystal structure of $Ca_5(PO_4)_2SiO_4$ silico-carnotite, *Tschermaks Mineral. Petrogr. Mitt.*, 16, 1, 1971.
32. Dickens, B., Schroeder, L. W., and Brown, W. E., Crystallographic studies of the role of Mg as a stabilizing impurity in β-$Ca_3(PO_4)_2$, *J. Solid State Chem.*, 10, 232, 1974.
33. Driessens, F. C. M., Physico-chemical interaction between biominerals and their environment, *Ber. Bunsenges. Physik, Chem.*, 82, 312, 1978.
34. Driessens, F. C. M., van Dijk, J. W. E. and Borggreven, J. M. P. M., Empirical border lines of the apatite stability field in the system $CaO-P_2O_5-H_2O-Na_2O-CO_2-MX$ at room temperature, *Z. Naturforsch.*, 34c, 165, 1979.
35. Driessens, F. C. M., The mineral in bone, dentin and tooth enamel, *Bull. Soc. Chim. Belg.*, 89, 663, 1980.
36. Driessens, F. C. M., Probable phase composition of the mineral in bone, *Z. Naturforsch.*, 35c, 357, 1980.
37. Driessens, F. C. M., van Dijk, J. W. E., Borggreven, J. M. P. M., and Verbeeck, R. M. H., Contribution to the physicochemical rationale for the caries reducing effect of fluoride, *J. Biol. Buccale*, 8, 239, 1980.
38. Driessens, F. C. M. and Verbeeck, R. M. H., Evidence for intermediate metastable states during equilibration of bone and dental tissues, *Z. Naturforsch.*, 35c, 262, 1980.
39. Driessens, F. C. M., Schaeken, H. G., Heyligers, H. J. M., and Verbeeck, R. M. H., Some physical properties of synthetic Na- and CO_3 containing apatites, in press.
40. Driessens, F. C. M. and Verbeeck, R. M. H., On the possible interaction mechanisms of strontium with sound and carious human tooth enamel, in press.
41. Driessens, F. C. M. and Verbeeck, R. M. H., Strontium metabolism and the interaction of strontium ions with bone mineral, in press.
42. Driessens, F. C. M. and Verbeeck, R. M. H., Probable phase composition of the mineral in sound enamel and dentine, in press.
43. Duff, E. J., Formation of apatites from calcium phosphates in potassium fluoride solutions, *J. Inorg. Nucl. Chem.*, 32, 3707, 1970.
44. Duff, E. J., The transformation brushite→fluorapatite and monetite→fluorapatite in aqueous potassium fluoride solution, *J. Chem. Soc. London*, 33, 1971.

45. Duff, E. J. and Grant, A. A., The preparation of phosphate ceramics for implantation, Abstract, Bioceramics Symposium, Sponsored by the European Society of Biomaterials, Keele, Canada, 1978.
46. Eanes, E. D., Gillesen, I. H., and Posner, A. S., Intermediate states in the precipitation of hydroxyapatite, *Nature (London)*, 208, 365, 1965.
47. Eanes, E. D. and Meyer, J. L., The influence of fluoride on apatite formation from unstable supersaturated solutions at pH 7.4, *J. Dent. Res.*, 57, 617, 1978.
48. Elliott, J. C., The Crystallographic Structure of Dental Enamel and Related Apatites, Ph.D. thesis, University of London, 1964.
49. Elliott, J. C., The interpretation of the infrared absorption spectra of some carbonate containing apatites, in *Tooth Enamel*, Stack, M. V. and Fearnhead, R. W., Eds., J. Wright & Sons, Bristol, England, 1965, 20.
50. Ericsson, Y., Tooth enamel solubility, *Acta Odont. Scand.*, 8(Supp.), 3, 1949.
51. Eysel, W. and Roy, D. M., Topotactic reaction of aragonite to hydroxyapatite, *Z. Kristallogr. Mineral.*, 141, 11, 1975.
52. Eysel, W., personal communication, 1980.
53. Farr, T. D. and Elmore, K. L., System $CaO-P_2O_5-HF-H_2O$: thermodynamic properties, *J. Am. Chem. Soc.*, 66, 315, 1962.
54. Fitton-Jackson, S., Randall, J. T., Wolstenholme, G. E. W., and O'Connor, C. M., Eds., *Bone Structure and Metabolism*, Little, Brown, Boston, 1956, 47.
55. Fix, W., Heymann, H., and Heinke, R., Subsolidus relations in the system $2CaO \cdot SiO_2-3\,CaO \cdot P_2O_5$, *J. Am. Ceram. Soc.*, 52, 346, 1969.
56. Fix, W., Wiemer, H. E., and Tolkemit, H. J., Das Zustandsschaubild $MnO_u-CaO-P_2O_5$ an Luft bei Temperaturen von 1510 bis 1610°C, *Arch. Eisenheuttenwes.*, 40, 217, 1969.
57. Forbes, J. E., Solubility of bone in solutions of magnesium salts, *J. Biol. Chem.*, 93, 255, 1931.
58. Forster, O., *Z. Anorg. Allg. Chem.*, 13, 1892, cited by Schleede, A., Schmidt, W., and Kindt, H., *Z. Elektrochem.*, 38, 633, 1932.
59. Fouretier, G., La precipitation du phosphate tricalcique et l'hydroxyapatite, *CR Seances Hebd. Acad. Sci.(Paris)*, 205, 413, 1937.
60. Frondel, C., Whitlockite a new calcium phosphate $Ca_3(PO_4)_2$, *Am. Mineral.*, 26, 145, 1941.
61. Garrels, R. M. and Christ, C. L., *Solutions, Minerals and Equilibria*, Harper and Row, New York, 1965.
62. Gonzalez Diaz, P. F. and Santos, M., The hydrogen motion in non-stoichiometric hydroxyapatites, *J. Solid. State Chem.*, 23, 265, 1978.
63. Gopal, R., Calvo, C., Ito, J., and Sabine, W. K., Crystal Structure of synthetic Mg-whitlockite $Ca_{18}Mg_2H_2(PO_4)_{14}$, *Can. J. Chem.*, 52, 1155, 1974.
64. Gottschall, A. J., Heats of formation of hydroxy-, fluor-, and chlorapatites, *J. S. Afr. Chem. Inst.*, 11, 45, 1958.
65. Greenwald, I., The effect of phosphate on the solubility of calcium carbonate and of bicarbonate on the solubility of calcium and magnesium phosphate, *J. Biol. Chem.*, 161, 697, 1945.
66. Gregory, T. M., Moreno, E. C., and Brown, W. E., Solubility of $CaHPO_4 \cdot 2H_2O$ in the system $Ca(OH)^2-H_3PO_4-H_2O$ at 5, 15, 25 and 37°C, *J. Res. Natl. Bur. Stand.*, 74A, 461, 1970.
67. Gregory, T. M., Moreno, E. C., Patel, J. M., and Brown, W. E., Solubility of β-$Ca_3(PO_4)_2$ in the system $Ca(OH)_2-H_3PO_4-H_2O$ at 5, 15, 25 and 37°C, *J. Res. Natl. Bur. Stand.*, 78A, 667, 1974.
68. Grøn, P., Spinelli, M., Trautz, O., and Brudevold, F., The effect of carbonate on the solubility of hydroxylapatite, *Arch. Oral Biol.*, 8, 251, 1963.
69. Gullbrandsen, R. A. and Roberson, C. E., Inorganic phosphorus in seawater, in *Environmental Phosphorus Handbook*, Griffith, E. C., Beeton, A., Spencer, J. M., and Mitchell, D. T., Eds., John Wiley & Sons, New York, 1973, 121.
70. Hägele, G. and Machatschki, M., Britolith, ein Cerium-erde Silikat-Apatit, *Naturwischaften*, 27, 132, 1939.
71. Hagen, A. R., Studies on fluorapatite. II. The solubility behavior, *J. Dent. Res.*, 54, 384, 1975.
72. Harper, R. A. and Posner, A. S., Measurement of non-crystalline calcium phosphate in bone mineral, *Proc. Soc. Exp. Biol. Med.*, 122, 137, 1966.
73. Harth, R., Thesis, Technische Hochschule, Munich, 1961.
74. Hayek, E. and Newesely, H., Ueber die Existenz von Tricalciumphosphat in wässriger Lösung, *Monatsh Chem.*, 89, 88, 1958.
75. Hayek, E., Konetschny, H., and Schnell, E., Synthese und Formel des biologischen und des mineralischen Carbonat-Apatits, *Angew. Chem.*, 5, 679, 1966.
76. Heughebaert, J. C. and Montel, G., Sur L'existence d'une série de solides de composition variable. correspondant au phosphate tricalcique precipité, *CR Acad. Sci. Ser C:*, 270, 1585, 1970.
77. Heyligers, H. J. M., Driessens, F. C. M., and Verbeeck, R. M. H., Lattice parameters and cation distribution of solid solutions of calcium and strontium hydroxy-apatite, *Calcif. Tissue Int.*, 29, 127, 1979.

78. Hill, W. L., Faust, G. T., and Reynolds, D. S., The binary system P_2O_5−$2CaO \cdot P_2O_5$ *Am. J. Sci.*, 242, 542, 1944.
79. Holt, L. E., Lamer, V. K., and Chown, H. B., The solubility product of secondary and tertiary calcium phosphate under various conditions, *J. Biol. Chem.*, 64, 509, 1925.
80. Ito, J., Synthesis of cerite, *J. Res. Natl. Bur. Stand.*, 72A, 355, 1968.
81. Jackson, S. A., Cartwright, A. G., and Lewis, D., The morphology of bone mineral crystals, *Calcif. Tissue Res.*, 25, 217, 1978.
82. De Jong, W. F., La substance minérale dans les os, *Rec. Trav. Chim. Pays-Bas*, 45, 445, 1926.
83. Kaufmann, H. W. and Kleinberg, I., An X-ray crystallographic examination of calcium phosphate formation in $Ca(OH)_2/H_3PO_4$ mixtures, *Calcif. Tissue Res.*, 22, 253, 1977.
84. Kaufmann, H. W. and Kleinberg, I., Studies on the incongruent solubility of hydroxyapatite, *Calcif. Tissue Int.*, 27, 143, 1979.
85. Keppler, U., Zum Whitlockit-Problem, *Neues Jahrb. Mineral. Monatsh.*, 6, 171, 1965.
86. Keppler, U., Structural investigation of calcium phosphate and isotypic structures, *Bull. Soc. Chim. Fr.*, 1771, 1968.
87. Kibby, C. L. and Hall, W. K., *The Chemistry of Biosurfaces*, Vol. 2, Hair, M. L., Ed., Marcel Dekker, New York, 1972, 663.
88. Kleinmann, H., Die Zustandsform des Calciums in Geweben und Gewebssäften, *Biochem. Z.*, 196, 98, 1928.
89. Klement, R. and Trömel, G., Hydroxyapatit der Hauptkomponent der anorganischen Substanz in Knochen und Zähne, *Z. Physiol. Chem.*, 213, 263, 1932.
90. Klement, R., Isomorphe Substitution von Phosphor in Apatite durch Silicium und Schwefel, *Naturwissenschaften*, 27, 57, 1939.
91. Klement, R. and Haselbeck, H., Apatite und Wagnerite von zweiwertigen Metalle, *Z. Anorg. Chem.*, 336, 113, 1965.
92. Klement, R. and Steckenreiter, F., Untersuchungen über isomorphen Ersatz der Elemente in Alkali-Erdalkaliphosphaten, *Z. Anorg. Allg. Chem.*, 245, 236, 1940.
93. Kreidler, E. R. and Hummel, F. A., Phase relationships in the system SrO−P_2O_5 and the influence of water vapor on the formation of $Sr_4P_2O_9$, *Inorg. Chem.*, 6, 884, 1967.
94. Kühl, G. and Nebergall, W. H., Hydrogenphosphat- und carbonatapatite, *Z. Anorg. Allg. Chem.*, 324, 313, 1963.
95. Kuyper, A. C., The chemistry of bone formation. II. Some factors affect the solubility of calcium phosphate in blood serum, *J. Biol. Chem.*, 159, 417, 1945.
96. Kuyper, A. C., The chemistry of bone formation. I. The composition of precipitates formed from salt solutions, *J. Biol. Chem.*, 159, 411, 1945b.
97. Labarthe, J. C., Bonel, G., and Montel, G., Sur la localisation des ions carbonate dans le réseau des apatites calcique, *CR Acad. Sci. Ser. C:*, 273, 349, 1971.
98. Labarthe, J. C., Therasse, M., Bonel, G., and Montel, G., Sur la structure des apatites phosphocalcique carbonatées de type B, *CR Acad. Sci. Ser. C:*, 276, 1175, 1973.
99. Labarthe, J. C., Bonel, G., and Montel, G., Sur la structure et les proprietes des apatites carbonatées de type B phosphocalcique, *Ann. Chim.*, 8, 289, 1973.
100. LeGeros, R. Z., Crystallographic Studies of the Carbonate Substitution in the Apatite Structure, Ph.D. thesis, New York University, 1967.
101. LeGeros, R. Z., Trautz, O. R., LeGeros, J. P., and Klein, E., Apatite crystallites: effects of carbonate on morphology, *Science*, 155, 1409, 1967.
102. LeGeros, R. Z., LeGeros, J. P., Trautz, O. R., and Klein, E., Spectral properties of carbonate in carbonate-containing apatites, *Dev. Appl. Spectrosc.*, 7B, 3, 1970.
103. LeGeros, R. Z., Shirra, W. P., Miravite, M. A., and LeGeros, J. P., Amorphous calcium phosphate synthetic and biological, *Colloq. Int. CNRS*, 230, 105, 1975.
104. Logan, M. A. and Taylor, H. L., Solubility of bone salts, *J. Biol. Chem.*, 119, 293, 1937.
105. Macaskill, J. B. and Bates, R. G., Solubility product constant of calcium fluoride, *J. Phys. Chem.*, 81, 496, 1977.
106. McCann, H. G., The solubility of fluorapatite and its relationship to that of calcium fluoride, *Arch. Oral Biol.*, 13, 987, 1968.
107. McCauley, R. A., Hummel, F. A., and Hoffman, M. V., Phase equilibria and Eu, Tb, and Mn-activated luminescent phases in the CaO−MgO−P_2O_5 system, *J. Electrochem. Soc.*, 118, 755, 1971.
108. McConnel, D., Substitution of SiO_4 and SO_4 groups for PO_4 groups in the apatite structure: ellestadite, the end member, *Am. Mineral.*, 22, 977, 1937.
109. McDowell, H., Brown, W. E., and Sutter, J. R., Solubility study of calcium hydrogen phosphate, ion-pair formation, *Inorg. Chem.*, 10, 1638, 1971.
110. Meyer, J. L. and Eanes, E. D., A thermodynamic analysis of the amorphous to crystalline calcium phosphate transformation, *Calcif. Tissue Res.*, 25, 59, 1978.

111. Meyer, J. L. and Eanes, E. D., A thermodynamic analysis of the secondary transition in the spontaneous precipitation of calcium phosphate, *Calcif. Tissue Res.*, 25, 209, 1978b.
112. Mikhailov, A. S., Calculation of the free energies of formation and the solubility products of certain phosphates of the apatite group, *Russ. J. Phys. Chem.*, 41, 461, 1967.
113. Montel, G., Méchanismes de la synthése de l'apatite de fluor, *Ann. Chim.*, 3, 313, 1958.
114. Moreno, E. C., Brown, W. E., and Osborn, G., Stability of dicalcium phosphate dihydrate in aqueous solutions and solubility of octocalcium phosphate, *Proc. Soil Sci. Soc. Am.*, 24, 99, 1960.
115. Moreno, E. C., Gregory, T. M., and Brown, W. E., Preparation and solubility of hydroxyapatite, *J. Res. Natl. Bur. Stand.*, 72A, 773, 1968.
116. Moreno, E. C., Kresak, M., and Zahradnik, R. T., Fluoridated hydroxyapatite solubility and caries formation, *Nature (London)*, 247, 64, 1974.
117. Moreno, E. C., Kresak, M., and Zahradnik, R. T., Physicochemical aspects of fluoride apatite systems relevant to the study of dental caries, *Caries Res.*, 11(Suppl. 1), 142, 1977.
118. Nacken, R., Das System CaF_2–$Ca_{10}(PO_4)_6F_2$, *Zentralbl. Mineral. Geol.*, 545, 1912.
119. Narasaraju, T. S. B. and Rao, V. L. N., A new interpretation of the solubility equilibria of hydroxyapatite, *Z. Physik. Chem.*, 255, 655, 1974.
120. Nayak, N. and Chiranjeevi Rao, S. V., pH dependence of solubility of calcium-cadmium hydroxylapatites, *J. Indian Chem. Soc.*, 52, 903, 1975.
121. Neuman, W. F., Toribara, T. Y., and Mulryan, B. J., The surface chemistry of bone. IX. Carbonate/phosphate exchange, *J. Am. Chem. Soc.*, 78, 4263, 1956.
122. Neuman, W. F. and Mulryan, B. J., Synthetic hydroxyapatite crystals. IV. Magnesium incorporation, *Calcif. Tissue Res.*, 7, 133, 1971.
123. Newesely, H., Darstellung von Oktacalciumphosphat durch homogene Kristallisation, *Monatsh. Chem.*, 91, 1020, 1960.
124. Newesely, H., Changes in crystal types of low solubility calcium phosphates in the presence of accompanying ions, *Arch. Oral Biol.*, 6(suppl.), 174, 1961.
125. Newesely, H., Kristallchemische und mikromorphologische Untersuchungen an Carbonat-Apatiten, *Monatsh. Chem.*, 94, 270, 1963.
126. Newesely, H., Umwandlungsvorgänge bei Calciumphosphaten, in *Calcified Tissues*, Fleisch, H., Blackwood, H. J. J., and Owen, M., Eds., Springer Verlag, Basel, 1965, 136.
127. Nurse, R. W., Welch, J. H., and Gutt, W., High-temperature phase equilibria in the system dicalcium silicate-tricalcium phosphate, *J. Chem. Soc.*, 1077, 1959.
128. Oelsen, W. and Maetz, H., Zur Metallurgie des Thomasverfahrens, *Arch. Eisenheuttenwes.*, 19, 111, 1948.
129. Osborn, J. F. and Newesely, H., The material science of calcium phosphate ceramics, *Biomaterials*, 1, 108, 1980.
130. Patel, P. R., Gregory, T. M., and Brown, W. E., Solubility of $CaHPO_4 \cdot 2H_2O$ in the quaternary system $Ca(OH)_2$–H_3PO_4–NaCl–H_2O at 25°C, *J. Res. Natl. Bur. Stand.*, 78A, 675, 1974.
131. Posner, A. S. and Stephenson, S. R., Isomorphous substitution in enamel apatite, *J. Am. Dent. Assoc.*, 46, 257, 1953.
132. Prener, J. S., The growth and crystallographic properties of calcium fluor- and chlorapatite crystals, *J. Electrochem. Soc.*, 114, 77, 1967.
133. Quinaux, N. and Richelle, L. J., X-ray diffraction and infrared analysis of bone specific gravity fractions in the growing rat, *Isr. J. Med. Sci.*, 3, 677, 1967.
134. Riboud, P. V., Dosage de l'eau dans l'hydroxyapatite à haute température *CR Acad. Sci. Ser. C:*, 269, 691, 1969.
135. Robinson, R. A. and Watson, M. L., Crystal-collagen relationship in bone as observed in the electron microscope, *Ann. NY Acad. Sci.*, 60, 596, 1955.
136. Romo, L. A., Synthesis of carbonato-apatite, *J. Am. Chem. Soc.*, 76, 3924, 1954.
137. Rootare, H. M., Deitz, V. R., and Carpenter, F. G., Solubility product phenomena in hydroxyapatite-water systems, *J. Colloid Sci.*, 17, 179, 1962.
138. Roseberry, H. H., Hastings, A. B., and Morse, J. K., X-ray analysis of bone and teeth, *J. Biol. Chem.*, 90, 395, 1931.
139. Rowles, S. L., The precipitation of whitlockite from aqueous solutions, *Bull. Soc. Chim. Fr.*, 1797, 1968.
140. Roy, D. M., Crystal growth of hydroxyapatite, *Mater. Res. Bull.*, 6, 1337, 1971.
141. Roy, D. M. and Linnehan, S. K., Hydroxyapatite formed from coral skeletal carbonate by hydrothermal exchange, *Nature (London)*, 247, 220, 1974.
142. Sanfourche, M. A., Recherches sur l'acide phosphorique et les phosphates. I. La formation des phosphates alcalino-terreux basique, *Bull. Soc. Chim. Fr.*, 53, 951, 1933.
143. Sata, T., Phase relationship in the system $3CaO \cdot P_2O_5$–$MgO \cdot SiO_2$–SiO_2, *Bull. Chem. Soc. Jpn.*, 31, 408, 1958.

144. **Sarver, J. F., Hoffman, M. V., and Hummel, F. A.**, Phase equilibria and tin-activated luminescence in strontium orthophosphate systems, *J. Electrochem. Soc.*, 108, 1103, 1961.
145. **Schaeken, H. G., Verbeeck, R. M. H., Driessens, F. C. M., and Thun, H. P.**, The variation of lattice parameters with composition in solid solutions of hydroxylapatite and fluorapatite, *Bull. Soc. Chim. Belg.*, 84, 881, 1975.
146. **Schaeken, H. G. and Driessens, F. C. M.**, unpublished results, 1981.
147. **Schaeken, H. G., Driessens, F. C. M., and Verbeeck, R. M. H.**, Phase diagram of the Ca-rich side of the system $CaO-P_2O_5-H_2O-CO_2-Na_2O$ at 870°C under nearly dry CO_2, in press.
148. **Schleede, A., Schmidt, W., and Kindt, H.**, Zur Kenntnis der Calciumphosphate und Apatite, *Z. Elektrochem.*, 38, 633, 1932.
149. **Sendroy, J. and Hastings, A. B.**, Studies on the solubility of calcium salts, *J. Biol. Chem.*, 71, 783, 1927.
150. **Seuter, A. M. J. H.**, Existence region of calcium hydroxylapatite, in *Reactivity of Solids*, Anderson, J. S., Roberts, M. W., and Stone, F. S., Eds., Chapman & Hall, London, 1972, 896.
151. **Simpson, D. R.**, The nature of alkali carbonate apatites, *Am. Mineral.*, 49, 363, 1964.
152. **Simpson, D. R.**, Effect of pH and solution concentration on the composition of carbonate apatite, *Am. Mineral.*, 52, 896, 1967.
153. **Simpson, D. R.**, Substitution in apatite. II. Low-temperature fluoride-hydroxyl apatite, *Am. Mineral.*, 53, 1953, 1968.
154. **Skinner, H. C. W.**, Phase relations in the $CaO-P_2O_5-H_2O$ system from 300 to 600°C at 2 kb H_2O pressure, *Am. J. Sci.*, 273, 545, 1973.
155. **Stone, P. E., Egan, E. P., and Lehr, J. R.**, Phase relations in the system $CaO-Al_2O_3-P_2O_5-$, *J. Am. Ceram. Soc.*, 39, 89, 1956.
156. **St. Pierre, P. D. S.**, Constitution of bone china. I. High-temperature phase equilibrium studies in the system tricalciumphosphate alumina-silica, *J. Am. Ceram. Soc.*, 37, 243, 1954.
157. **St. Pierre, P. D. S.**, Constitution of bone china. III. High-temperature phase equilibrium studies in the system tricalciumphosphate-anorthite-silica *J. Am. Ceram. Soc.*, 39, 147, 1956.
158. **Strates, B. S., Neuman, W. F., and Levinskas, G. J.**, The solubility of bone mineral. II. Precipitation of near-neutral solutions of calcium and phosphate, *J. Phys. Chem.*, 61, 279, 1957.
159. **Tagai, H. and Aoki, H.**, Preparation of synthetic hydroxyapatite and sintering of apatite ceramics, Abstract, Bioceramics Symposium, sponsored by the European Society of Biomaterials, Keele, Canada, 1978.
160. **Termine, J. D. and Posner, A. S.**, Amorphous/crystalline interrelationships in bone mineral, *Calcif. Tissue Res.*, 1, 8, 1967.
161. **Termine, J. D. and Posner, A. S.**, Calcium phosphate formation in vitro. I. Factors affecting initial phase separation, *Arch. Biochem. Biophys.*, 140, 307, 1970.
162. **Termine, J. D., Peckauskas, R. A., and Posner, A. S.**, Calcium phosphate formation in vitro. II. Effects of environment on amorphous-crystalline transformation, *Arch. Biochem. Biophys.*, 140, 318, 1970.
163. **Termine, J. D. and Eanes, E. D.**, Comparative chemistry of amorphous and apatitic calcium phosphate preparations, *Calcif. Tissue Res.*, 10, 171, 1972.
164. **Termine, J. D., Eanes, E. D., Greenfield, D. J., Nylen, M. U., and Harper, R. A.**, Hydrazinedeproteinated bone mineral: physical and chemical properties, *Calcif. Tissue Res.*, 12, 73, 1973.
165. **Tovborg Jensen, A. and Rowles, S. L.**, Lattice constants and magnesium contents of some naturally occurring whitlockites, *Nature (London)*, 179, 912, 1057.
166. **Trautz, O. R.**, X-ray diffraction of biological and synthetic apatites, *Ann. N.Y. Acad. Sci.*, 60, 696, 1955.
167. **Trömel, G.**, Untersuchungen über die Bildung eines halogenfreien Apatis aus basischen Calciumphosphaten, *Z. Physik. Chem.*, 158, 422, 1932.
168. **Trömel, G. and Möller, H.**, Die Bildung schwer löslicher Calciumphosphate aus wässriger Lösung und die Beziehungen dieser Phosphate zur Apatitgruppe, *Z. Anorg. Allg. Chem.*, 206, 227, 1932.
169. **Trömel, G.**, Untersuchungen des ternären Systems $CaO-P_2O_5-SiO_2$ und die Bedeutung in den praktischen Gebrauch von basichen Bessemer Schlacken, *Stahl Eisen*, 63, 21, 1943.
170. **Trömel, G. and Fix, W.**, Untersuchungen im System Kalk-Phosphorsäure, *Arch. Eisenheuttenwes.*, 32, 209, 1961.
171. **Trömel, G., Fix, W., and Fritze, H. W.**, Zusammenfassende Darstellung der Gleichgewichte Zwischen Eisen and Kalkhaltigen Phosphatschlacken, *Arch. Eisenheuttenwes.*, 32, 353, 1961.
172. **Trombe, J. C. and Montel, G.**, Conditions de la synthèse d'une apatite nouvelle contenant d'ions de sulfur, *CR Acad. Sci. Ser. C:*, 280, 567, 1975.
173. **Trombe, J. C., Szilagyi, J., and Montel, G.**, Sur l'existence d'un phosphate tricalcique anhydre β, contentant des ions sulfate en solution solide, *CR Acad. Sci. Ser. C;* 283, 199, 1976.
174. **Verbeeck, R. M. H., Hauben, M., Thun, H. P., and Verbeek, F.**, Solubility and solution behaviour of strontiumhydroxyapatite, *Z. Physik. Chem.*, 108, 203, 1977.

175. Verbeeck, R. M. H., Khan, R. A., and Thun, H. P., Potentiometric determination of the thermodynamic solubility product of calcium fluoride, *Bull. Soc. Chim. Belg.*, 86, 503, 1977.
176. Verbeeck, R. M. H., Thun, H. P., and Driessens, F. C. M., Solubility behaviour of defective fluorhydroxyapatites, *Ber. Bunsenges. Physik. Chem.*, 84, 159, 1980.
177. Verbeeck, R. M. H., Khan, R. A., Thun, H. P., and Driessens, F. C. M., Solubility of thermally synthesized fluorhydroxyapatites, *Bull. Soc. Chim. Belg.*, 88, 751, 1979.
178. Verbeeck, R. M. H., Oplosbaarheidsgedrag van Calcium en Strontium Bevattende Hydroxy-Apatieten, Thesis, University of Ghent, Ghent, Belgium, 1980.
179. Verbeeck, R. M. H., Heyligers, H. J. M., Driessens, F. C. M., and Schaeken, H. G., Effect of dehydration of calcium hydroxyapatite on its cell parameters, *Z. Anorg. Allg. Chem.*, 466, 1980.
180. Verbeeck, R. M. H. and Driessens, F. C. M., Metastable states in calcium phosphate-aqueous phase equilibrations, *J. Crystal Growth*, 53, 55, 1981.
181. Verbeeck, R. M. H., Kiekens, P., and Driessens, F. C. M., Phase equilibria in strontium orthophosphate solutions at 25°C., *Z. Physik. Chem.*, 124, 45, 1981.
182. Verbeeck, R. M. H., Lassuyt, C. J., Heyligers, H. J. M., Driessens, F. C. M., and Vrolijk, J. W. G. A., Lattice parameters and cation distribution of solid solutions of calcium and lead hydroxyapatite, *Calcif. Tissue Int.*, 33, 243, 1981.
183. Verbeeck, R. M. H., Thun, H. P., and Verbeek, F., Solubility behaviour of synthetic Ca-Sr hydroxyapatites, *Z. Physik. Chem.*, in press.
184. Vignoles, C., Bonel, G. and Montel, G., Sur la composition des apatites carbonatées de type B préparée en milieu sodé, *CR Acad. Sci. Ser. C:*, 280, 361, 1975.
185. Wallaeys, R. and Chaudron, G., Sur la préparation de certaines apatites mixtes, *CR Acad. Sci. Ser. C:*, 231, 355, 1950.
186. Wallaeys, R., Apatites phosphocalcique, *Ann. Chim.*, 7, 808, 1952.
187. Welch, J. H. and Gutt, W., High-temperature studies of the system calciumoxyde-phosphorus pentoxide, *J. Chem. Soc.*, 4442, 1961.
188. Wendt, G. L. and Clarke, A. H., An electrometric study of the neutralization of phosphoric acid by calcium hydroxyde, *J. Am. Chem. Soc.*, 45, 881, 1923.
189. Wier, D. R., Chien, S. H., and Black, C. A., Solubility of hydroxyapatite, *Soil Sci.*, 111, 107, 1971.
190. Wier, D. R., Chien, S. H., and Black, C. A., Transformation of hydroxyapatite to fluor-apatite, *Soil Sci. Soc. Am. Proc.*, 36, 285, 1972.
191. De With, G., van Dijk, H. J. A., Hattu, N., and Prijs, K., Preparation, microstructure and mechanical properties of dense polycrystalline hydroxyapatite, *J. Mater. Sci.*, in press.
192. Zahradnik, R., Rericha, R., Axamit, P., Rezabkova, M., and Skramovsky, S., Ueber die Reaktion einiger Kationen von Schwermetallen mit wenig löslichen Calciumverbindungen, *Coll. Czech. Chem. Commun.*, 25, 146, 1960.
193. Zapanta LeGeros, R., Effect of carbonate on the lattice parameters of apatite, *Nature (London)*, 206, 403, 1965.

Chapter 2

FORMATION AND DECAY OF THE TOOTH

J. H. M. Wöltgens

TABLE OF CONTENTS

I. MINERALIZATION IN DEVELOPING TEETH

The development of the tooth germ from soft tissue to the most hard tissue in the organism occurs according to a very fascinating, well-programed scheme. The tooth originates from an interaction between epithelial and ectomesenchymal cells which differentiate into an enamel organ and a dental papilla. After further differentiation, the enamel organ is bell-shaped and surrounded by an inner and outer enamel epithelium. In between, the stellate reticulum and the stratum intermedium are formed. This peel form is induced by the ectomesenchyme cells, which differentiate towards the odontoblasts.[23] The rest of the dental papilla develops into the subodontoblastic layer and pulp cells (Figure 1).

Circumstantial evidence supports the view that collagen plays a crucial factor in regulating the odontogenesis. This has been shown by cultivation of tooth germs in vitamin C-deficient medium,[35] resulting in a stop in differentiation. In this respect, it is well-known that ascorbic acid is needed for the normal formation of collagen. At the same time, dedifferentiation of ameloblasts is observed together with a degeneration of the odontoblasts as the consequence of impaired collagen production (Figure 2). Apart from that, the collagen also serves as matrix (predentin) along which mineralization of dentin takes place. This mineralization always starts several microns' distance from the odontoblasts. This growth occurs from nucleation centers of calcium phosphate salts, which have already been formed in extracellular vesicles derived from the extensions of the odontoblasts. At that time the ameloblasts secrete enamel protein in the so-called Tomes process, which contains calcium and immediately acts as the nucleation center directly outside the cells. In contrast to dentinogenesis, during the Tomes process no nuclei formation of the calcium salt by extracellular vesicles of the ameloblasts is found. It has been suggested that the calcium salts in the already-formed dentin serve as nuclei for the mineralization of the enamel.[7]

After this initial mineralization of the enamel, the appearance and function of the ameloblasts change from secretory towards a resorptive cell (Figure 3). It seems that parts of the protein already secreted in the enamel and called amelogenin changes from large molecules toward at least two smaller molecules mainly, during the resorptive phase of the ameloblasts. When the provisional mineralized enamel reaches its full thickness, a sudden further maturation of this provisional mineralized enamel takes place starting from the enamel dentin junction.

It is reasonable to suggest that the structure of the mineralized enamel or the dentin is mainly determined by the ground matrix or ground substance of the uncalcified predentin or pre-enamel. However, in both tissues the main calcium salt is hydroxyapatite $Ca_{10}(PO_4)_6(OH)_2$. The mineralization of the predentin or the pre-enamel occurs between the ameloblastic layer of the enamel organ and the odontoblastic layer of the dental papilla. The main difference between the dentin and the enamel after the formation is the fact that odontoblasts remain in contact with the enamel-dentin junction with their extensions, whereas the enamel after the disappearance of the Tomes process possesses no cellular contact with the dentin-enamel junction. Another difference is that dentin formation occurs in discrete additional layers, whereas enamel maturation starts diffusely from the enamel-dentin junction after the calcium-containing enamel proteins have reached their full length. At that time the internal and external ameloblastic layers reach each other, disappear, and the tooth erupts.

During the mineralization of the tooth germs a very high alkaline phosphatase has been demonstrated in the subodontoblastic layer and the stratum intermedium, which is thought to stimulate the mineralization. It is hypothesized that the enzyme promotes mineralizations by accumulating calcium,[51] by supplying phosphates[53] and by eliminating inhibitors of mineralization such as inorganic pyrophosphate (PP_i).[70]

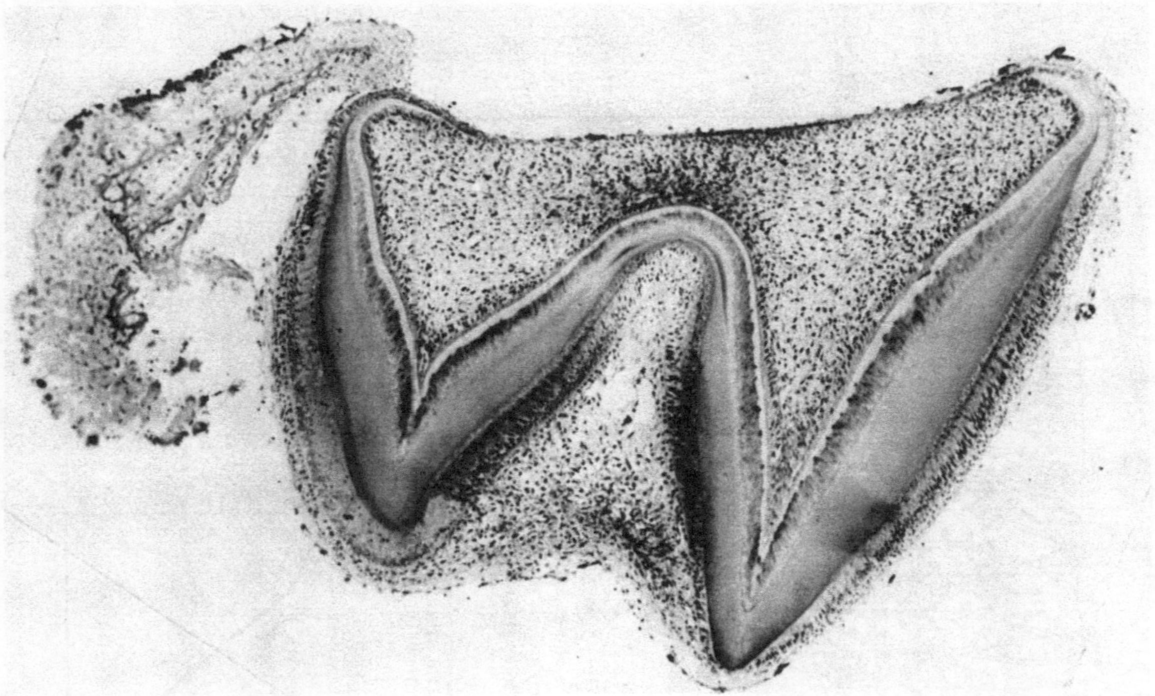

FIGURE 1. Early development stage of a hamster molar. 1. Ameloblasts. 2. Enamel matrix. 3. Dentin. 4. Odontoblasts. (Toludine blue staining; magnification × 100.)

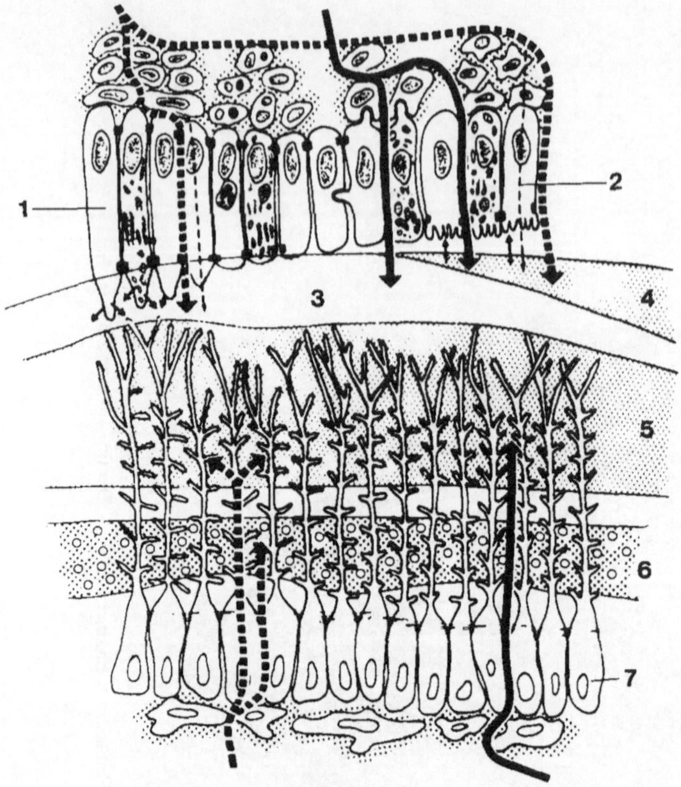

FIGURE 2. Ameloblasts and odontoblasts. 1. Secretory ameloblasts. 2. Resorptive ameloblasts. 3. Enamel matrix. 4. Mature enamel. 5. Dentin. 6. Predentin. 7. Odontoblasts. --- Ca pathway, —— Pi pathway.

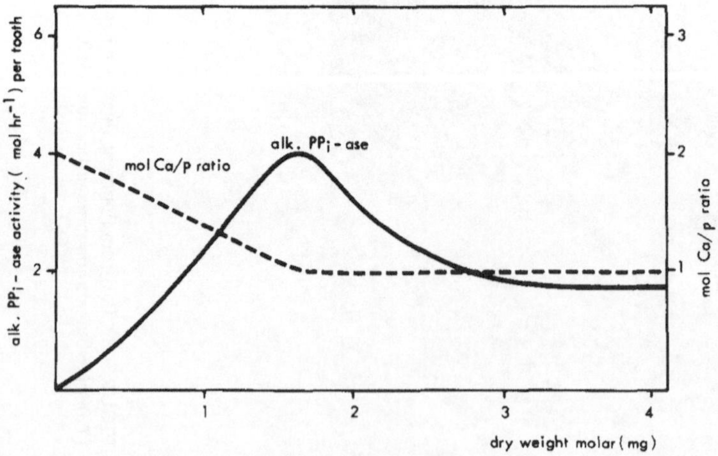

FIGURE 3. Relationship between alkaline phosphatase and the mol Ca/P ratio during the development of hamster molars. (From Wöltgens, J. H. M., *Ned. Tijdschr. Tandheelkd.*, 86, 332, 1979. With permission.)

The fact that in hamsters before the alkaline phosphatase increases, calcium already has been accumulated, which makes a function of alkaline phosphatase in calcium accumulation very unlikely. The phosphate increases, however, coinciding with the increase in enzyme activity. This leads to a decrease in mol Ca/P ratio from 4 to 2 (Figure 3). The high Ca/P ratio at the beginning of the mineralization can only be explained by the fact that in that period calcium is bound to another anion than phosphate. Therefore, the high alkaline phosphatase seems to be more correlated with the phosphate metabolism or with the phosphate accumulation.[70] The stimulation of the mineralization by the enzyme can be explained by supplying phosphate or eliminating inhibitors of the mineralization such as PP_i. With respect to the question whether the high enzyme activity is involved in initiating nucleation or is stimulating further growth of nucleation centers already present in the extracellular fluid, the following results have to be mentioned:

1. The highest alkaline phosphatase is located in the stratum intermedium and the subodontoblastic layer, whereas the ameloblasts and odontoblasts which are closer to the mineralization front contain only a little enzyme activity.
2. Before the phosphate increases together with the increase in alkaline phosphatase, already calcium accumulation has taken place.[70]

Therefore, from both observations it can be concluded that the alkaline phosphatase is mainly concerned with the further growth of nuclei already present extracellularly, by providing a phosphate gradient from the stratum intermedium or subodontoblastic layer between the cells towards the nucleation centers in the enamel or dentin.

With respect to the formation of the periodontal ligaments and cement, a short summary will be given. After the formation of enamel and dentin, the apical part of the external and internal ameloblastic layers fuse together and form the so-called sheath of Hertwig. The future cement-dentin junction is situated in between this layer and tooth papilla. The induction of the cement-forming cells from fibroblasts is the consequence of indirect contact between the fibroblasts and odontoblasts. This is possible after the disaggregation of the cells of the sheath of Hertwig. Collagen formation now starts between the bone and the cementum in the so-called periodontal ligament. This ligament is very important for keeping the tooth closely connected with the jaw.

The next section will go into further detail with respect to the enamel formation.

II. ENAMEL FORMATION

As described in the foregoing section the influence of two tissues, the epithelial ameloblasts and the ectomesenchymal odontoblast, on each other results in the formation first of dentin and then enamel. In this part, the development of the enamel will be discussed in more detail because changes in its composition or structure can have very dramatic consequences for the caries resistance of the tooth.

After differentiation of the internal ameloblastic layer, the ameloblast starts with the production of protein matrix for the enamel (Figure 1). The protein matrix is secreted in vesicles. At the same time, the ameloblast retracts from the dentin-enamel junction, forming the Tomes process. Immediately after secretion, the primary mineralization of the protein takes place. The crystals are located within a prisma almost perpendicular to the membranes of the Tomes processes. However, because the retraction of the ameloblasts is in an upward direction, the crystals are more densely packed in the head of the prisma than in a somewhat lower situated area (Figure 4). The retraction also leads to a deviation in the original hexagonal appearance of the cross-section of a prisma toward a more keyhole structure. One keyhole will be formed by

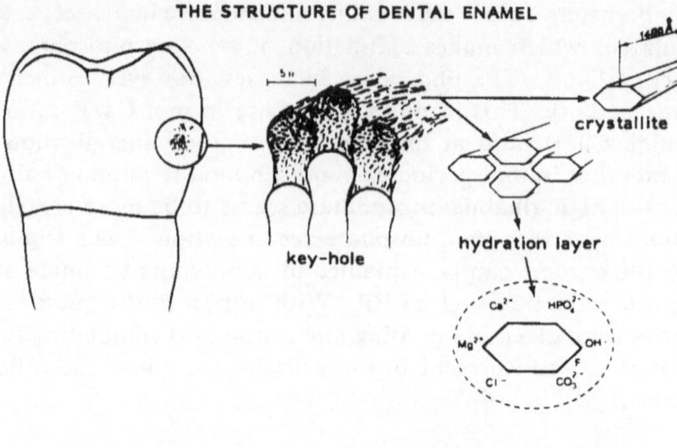

FIGURE 4. Microstructure of the dental enamel. (From Wöltgens, J. H. M., in *Preventieve Tandheelkunde,* Stafleu en Tholen, Houwink, 1979. With permission.)

one ameloblast. When the enamel matrix has been deposited over the whole enamel, the secreting ameloblast changes into a resorbing cell. These cells contain a ruffle at their distal ends during the last phase of enamel matrix maturation, only connected to each other by the tight junctions. During that period, the amount of enamel protein decreases[67] and changes in composition coinciding with a decrease in H_2O and an increase in mineral up to 86% vol. From the work of Daculsi and Kerebel[15] it can be concluded that during the primary mineralization the hexagonal crystal increases in length from 150 to 600 Å, whereas during the further mineralization an increase in thickness occurs from 45 Å to 260 Å, suggesting a two-stage process. At the end of the enamel formation, the enamel surface will be formed that is mostly prismless.[63] This enamel is formed by degenerating ameloblasts over a long period. During that time a calcium salt with a mol Ca/P ratio lower than apatite is formed. It is also clear that great structural differences exist between the surface layer which is approximately 30 μm thick and the more interior layers of the enamel. After eruption, the remnants of the ameloblasts form the first cuticle of the enamel in the mouth (the membrane of Nasmyth). The enamel formation takes place with alternating velocities leading to a periodic bending of the prisms that results in the so-called Retzius lines and seem to contain, depending on the metabolic state of the ameloblasts, apatite with low or high CO_3^{2-} content.[9,72] Biochemical changes can be found in protein together with an increase in thickness of the mineral. According to Lyaruu et al.[37] two types of enamel protein can be described: amelogenines which change during the further development of the enamel and enamelins, which are closely related to the enamel crystals and do not change during the development of enamel.

During maturation, amelogenin protein is suddenly changed from a high molecular weight to a lower-sized amelogenin. Whereas the amelogenins originally contain very high amounts of proline, glutamic acid, leucine, and histidine, during maturation a marked decrease of proline and histidine together with aspartic acid, alanine, valine, lysine, and arginic acid could be found. This is possibly due to the action of resorbing ameloblasts. The enamelin remains rich in glycine, serine, glutamic acid, and alanine. It seems likely that the changes in the amount and composition of the amelogenins coincide with the increase in thickness of the crystals from 45 to 260 Å. However, no ideas exist about the reason for these changes. The crystals do not show different diffraction patterns than apatite. This means that the first enamel crystal already contains the apatitic structure or that other phases disappear during the pretreatment of the young enamel, e.g., CO_3^{2-}-containing calcium salts.

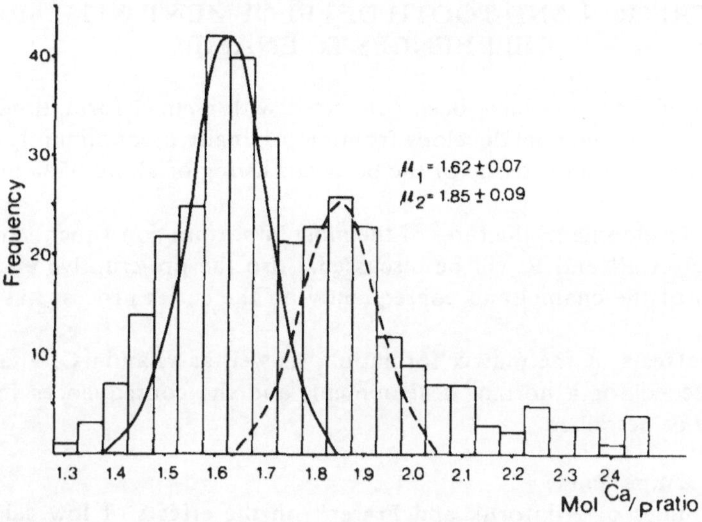

FIGURE 5. Gaussian fit to the frequency distribtuion of the molar Ca/P ratio in human enamel. (From Wöltgens, J. H. M. et al. *Arch. Oral Biol.*, 25, 435, 1980. With permission.)

As mentioned earlier, it has been found that during the development of molars of hamsters, the minerals change from a very high mol Ca^{2+}/P ratio toward more normal values of 1.2. Moreover in human enamel, two discrete calcium salts could be determined:[72] one with a mol Ca^{2+}/P ratio of 1.62 and the other with a mol Ca^{2+}/P ratio of 1.85 (Figure 5). Both findings suggest that apart from the normal apatite, Ca^{2+} has to be bound to another anion, which is masked by increasing amounts of apatite during the further development of the tooth. It seems from the study of Sobel and Hanok[60] that the mol Ca/P ratio of the enamel might be influenced by the diet given during the pre-eruptive development of the molar and that it coincides with a change in the $PO_4/2CO_3$ ratio of the serum. This again suggests that apart from apatite, other phases can exist in tooth enamel depending on the $PO_4/2CO_3$ ratio in the serum.

However, apart from the above-mentioned separate phases, in the human enamel surface a mol Ca^{2+}/P ratio lower than apatite exists that disappears during the posteruptive maturation (possibly whitlockite or octacalcium phosphate.[17] It is likely that differences in local environments in the enamel surface and in the more interior-situated layers have to be consequences of differences in the metabolic activities between the young and older, even degenerating, ameloblasts.

The function of the internal ameloblastic layer as well as the stratum intermedium or the stellate reticulum during the mineralization of the enamel still remains obscure. These cells seem to have different effects on the calcium and phosphate metabolism. Whereas the ameloblast will inhibit the influx of calcium in the enamel during the secretory phase, it does not influence the phosphate uptake.[6] It has been suggested that the phosphate uptake increases very dramatically in the stratum intermedium and the stellate reticulum by the presence of very high alkaline phosphatase and diffuses from there through the ameloblastic layer toward the mineralizing fronts.[71] In agreement with these findings are the results of Burger et al.[12] who showed that the matrix-containing vesicles in the ameloblasts do not contain Ca^{2+}. This also indicates that the mineralization of the enamel matrix takes place extracellularly. The maturation can be explained as a consequence of a further growth of nuclei formed extracellularly by a combined activity of a high alkaline phosphatase together with a decreased activity of the ameloblasts toward the efflux of Ca^{2+} from the mineralizing site of the enamel.

III. NUTRITION AND TOOTH DEVELOPMENT WITH SPECIAL REFERENCES TO ENAMEL

Many nutritional factors have been connected with enamel formation. As already mentioned, the dental enamel develops from an originally predominantly organic material toward the most hard tissue of the body consisting of about 98% inorganic material.

During this development, the fates of the most important constituents of the enamel mineral, namely Ca^{2+} and P will be discussed. Also the pre-eruptive effect of F^- on mineralization of the enamel and consequently on the caries process has to be taken into account.

Finally the effects of the matrix formation, as well as vitamin C, vitamin A, and vitamin D, needed for a normal development, and the consequences for the caries resistance, will be described.

A. Inorganic Components

From the studies of Nikiforuk and Fraser[46] on the effects of low calcium or low phosphate concentration in the serum of patients suffering from different diseases, it is clear that during the pre-eruptive mineralization of tooth enamel and dentin low Ca^{2+} leads to great disturbances in dental enamel. In contrast, the low phosphate concentrations in the serum lead to the impairment of the mineralization of bone and dentin.

The already described results of Sobel and Hanok[60] suggest that during Ca^{2+} deficiency, the Ca^{2+}/P ratio of the enamel will not markedly change, however, the $PO_4/2CO_3$ ratio will be low in contrast to the enamel from the low phosphate diet, in which the ratio is very high. From both experiments it can be concluded that Ca^{2+} is essential for the normal development of the enamel. The PO_4 concentration also seems to be influenced by the carbohydrate content in the blood. Carbohydrate increases the CO_3^{2-} content and diminishes the PO_4 concentrations leading to a lower $PO_4/2CO_3$ ratio which increases the caries susceptibility.

Closely related to the Ca and P metabolism are the vitamin D metabolites and the calcitonin hormone and parathyroid hormone (PTH). The general function of vitamin D is to facilitate normal mineralization of newly formed bone by keeping the levels of Ca and P sufficiently high to support normal mineralization.[27] Vitamin D deficiency during the pre-eruptive period of the tooth formation leads to hypoplasia of the dental enamel, although the enamel seems to be more resistant against the deficiency than bone or dentin possibly because vitamin D is more correlated with phosphate metabolism. The question that remains to be solved is whether this type of hypoplasia is more sensitive toward caries.

The same holds true for deficiencies of the calcium-regulating hormones. In this respect a very interesting result should be mentioned: calcitonin injected during the development of the teeth reduces dental caries.[66] On the other hand, other elements are very necessary for a normal development of the teeth. The most abundant trace element present in enamel is Mg^{2+}. Mg^{2+} as well as HCO_3^- in the enamel decreases the crystallinity of the mineral. Mg^{2+} is also needed as co-factor for very important metabolic processes so that a severe Mg^{2+} deficiency can result in complete atrophy of the developing tooth.

Another very important trace element is fluoride. Although there are some doubts about the pre-eruptive effects of F^- on caries reduction in man and animals, an inhibiting effect could be demonstrated.[63] On the other hand, the pre-eruptive effects of high concentrations during enamel formation are well-known as mottling or fluorosis. Several authors have suggested that fluoride acts on the secretory ameloblast leading to an abnormal matrix production.[33,34,39] Other authors suggest that F^- acts especially on the resorbing ameloblasts thereby influencing the matrix remaining behind in the definitive enamel.[20] Various authors have showed in this respect that F^- accumulation takes place before maturation of the dental enamel starts.[64,14,69] The various authors suggest that F^- is more related toward the enamel matrix than toward the inorganic part of the enamel. Because of this, the findings of Patterson et al.[47] have to be mentioned. Patterson demonstrated that fluoride changes the amino acid composition of the matrix. Comparing the changes with those in hamster amelogenesis during their development,[37] it is interesting to note that fluoride effects suggest a stimulation of the maturation of proteins. The maturation of the inorganic part of the enamel coincides with an absolute decrease in F^-. This suggests that abnormal crystals in fluorosis are not the result of a direct effect of F^-, but rather of an indirect effect of fluoride by changing the primary mineralization.

Angmar et al.[2] have demonstrated that plasma F^- peaks of about 10 μmol (0.2 ppm) can disturb amelogenesis and lead to highly porosed enamel, containing areas of hypocalcifications, especially along the enamel rods and cross-striations. The porosity is most pronounced in the subsurface. The matrix and crystal, changed by F^-, however, are more resistant to dissolution by acids of the microorganism. In the review of Navia[42] the effects of several elements on dental caries are summarized as follows:

Cariostatic: F^-, P.
Mild cariostatic: Mo, M_o, V, Cu, Sr, B, Li, Au.
Doubtful: Be, Co, Mn, Sn, Zn, Br, J.
Caries-inert: Ba, Al, Ni, Fe, Pd, Ti.
Caries-promoting elements: Se, Mg, Cd, Pt, Pb, Si.

However, as the author mentioned, it is not known whether these effects are brought about pre-eruptively or posteruptively.

B. Organic Components

The organic matrices in the dental enamel and dentin are produced by ameloblasts and odontoblasts, respectively. For the normal development of both cell types, two vitamins are very important: vitamin C and vitamin A. Although vitamin C is needed for collagen production by the mesenchymal cells, Bronckers et al.[11] demonstrated in an in vitro experiment that collagen formation is an absolute requisite for a normal development of the whole tooth. It seems that vitamin C acts indirectly on the amelogenesis by changing the inductive properties of the odontoblasts. The relationship between vitamin C and dental caries still remains obscure. In contrast to the vitamin C effects several authors[48,55,26] have demonstrated that vitamin A deficiency leads to an increase in caries susceptibility. In this respect Harris and Navia[26] suggest, based on the fact that the solubility of the dental enamel has not been changed, that vitamin A acts on dentin formation rather than on enamel. The influence of other vitamins are not that extensively investigated, so they will not be described here.

IV. POSTERUPTIVE MATURATION

When the crown has been fully mineralized, the tooth erupts by the formation of the root. The apex formation is the result of an interaction of the odontoblasts and fibroblasts of the bone. From epidemiological studies of Carlos and Gittelsohn[13] it is clear that after the eruption, the tooth enamel does not have its definitive strength and crystallinity as it does later. In this respect, they were able to show for each tooth an increase in caries incidence the first three years posteruptively. Thereafter, the caries

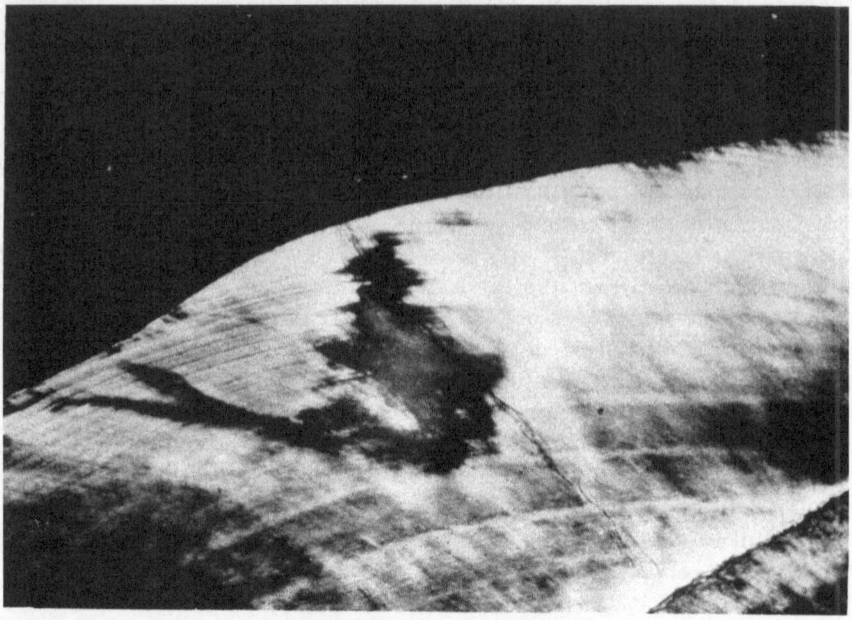

FIGURE 6. White spot formation.

incidence falls very sharply. This means that if the tooth has been attacked by acids within three years after eruption, it will be more easily dissolved by the acid. Apparently, the enamel will be changed after the eruption by the saliva and will become more resistant to caries. This process of maturation belongs not only to the young tooth, but also to older elements which have become more immature by the first acid dissolution. This incipient caries lesion is called "white spot" and is characterized by a relative intact enamel surface (Figure 6) and a demineralized subsurface. Backer Dirks[5] showed that such "immature" enamel can disappear later on. Comparing the distribution of the white spots in children at an age of 8 years and 15 years, Backer Dirks found that a remarkable amount of white spots changed toward sound enamel. From this study it can be concluded that partly demineralized enamel can be repaired by constituents of the saliva. Speirs[62] studied the processes of maturation and demineralization by coloring the hypomineralized areas of enamel with alizarin red S or naphtalene black in vitro. The progressive reduction in staining as a measure of maturation was greatest in the enamel that became the most caries-resistant. Indeed, for this maturation effect factors from the saliva seem to be essential because ligation of the salivary ducts retarded not only mineralization, but changed hypomineralized areas into incipient caries lesions.

Although it has been suggested that newly erupted molars contain calcium-deficient enamel, it seems that the addition of Ca^{2+} in the diet has a lesser effect on the maturation than phosphate does.[36] A finding of Speirs[62] should also be mentioned here: the fact that F^- given pre-eruptively leads to a highly hypomineralized tooth, whereas F^- stimulates mineralization given posteruptively. The above-mentioned publications suggest that during posteruptive development, changes occur in the dental enamel that coincide with a lower incidence of caries. However, because no quantitative chemical data exist on posteruptive maturation,[72] a technique was developed[72] with which it was possible to determine in very small parts of the enamel surface the Ca^{2+} and inorganic phosphate (Pi) content. With this method premolars with different posteruptive age extracted for orthodontic reasons were analyzed. The Ca^{2+} and phosphate contents in slices of the enamel cut from the labial surface of the human premolars both decrease from the surface to deeper layers, where they level off at a depth of about 60 μm in young as well as in older teeth. Posteruptive changes were found most clearly at a depth of about 25 μm. In these locations the Ca^{2+} and phosphate contents decrease with age. The change from a relatively low mol Ca/P ratio toward a higher mol Ca/P ratio has been interpreted in terms of calcium phosphate phases different from apatite and existing in the young enamel surface which disappear during maturation. Furthermore, it could be demonstrated that this maturation phenomenon coincides with a decreased loss of calcium and phosphate in the older dental enamel in artificial caries.[72] Again the highest change occurs at a depth of 25 to 30 μm from the surface.

The above-mentioned variations have not been found by electron microprobe analysis, possibly because in whole enamel they are masked by differences in density.[21] In earlier polarized light microscopical results,[71] it was demonstrated that the greatest part of this maturation takes place during the first year posteruptively. The same holds true for the process of remineralization. The increase in caries incidence of about three years, as has been shown by Carlos and Gittelsohn[13] can be explained by the assumption that there exists some delay of several years between the initial attack of the bacteria and the development of a full cavity, as registered by the dentist. From the findings of Carlos and Gittelsohn[13] it can be derived that the period of high caries activity takes place from six years when the first incisor or molar is erupting until three years after the eruption of the second molar (= 18 years). The third molars are not considered in this respect.

V. COMPOSITION OF ADULT TOOTH ENAMEL

If a fully formed dental enamel is examined, several microscopic structures can be discerned of which the Retzius lines as the consequence of different growth rate of the ameloblast are the most important. The Retzius lines reach the enamel surface, and there form perikymatas. Apart from these structures, the Hunter-Schreger bands can be seen as the result of the undulation of the ameloblasts, whereas after the eruption, the dental enamel rests of ameloblasts remain on the tooth as the so-called membrane of Nasmyth. The function of this membrane has to be clarified. The general chemical composition expressed as percentage of total dryweight is

Ca	P	Mg^{2+}	CO_2	Organic matter
36%	17%	0.4%	2.5%	1.3%

Although from the X-ray pattern of the dental enamel it is generally accepted that only one calcium salt exists in the enamel, evidence could be given for the existence of two separate phases of this structure: a calcium salt with a Ca/P ratio of 1.62 as well as a calcium salt with a Ca/P ratio of 1.85[72] From the cutting direction of the slices these authors concluded that the salts are present as alternating bands that run almost parallel to the surface. The salt with a Ca/P ratio of 1.85 might be a carbonate containing apatite that is more soluble and situated in the above-mentioned bands, e.g., the Retzius lines. In this way the histological appearance of the caries process extending along the Retzius lines or cross-striations could be explained by the preferential dissolution of these areas (Figure 6). In this respect, Boyde[9] suggests that in the core of an apatite crystal, a higher carbonate concentration also exists that should lead to the preferential dissolution of the core of the apatite crystal, comparable with the dissolution of a carbonate-rich synthetic apatite.[4] This contains about 5 to 10% of the CO_3^{2-} present in enamel. The other 90 to 95% CO_3^{2-} fills positions normally occupied by PO_4^{3-}. Electroneutrality can be obtained by void filling of a monovalent anion such as OH^- or F^-, or by a monovalent cation such as Na^+ or K^+. With respect to the organic matrix, it is clear that in adult tooth enamel the matrix becomes more insoluble.[18] The enamel surface differs in many ways from the more interior layers. Histologically, most of the outer part of the dental enamel (30 μm) appears to be prismless and is mostly located on the gingival third of the surface[63] which is built by the ameloblasts at the end of the crown formation. In these prismless zones, crystallites are located almost perpendicular to the surfaces and more closely and uniformly arranged than in the subsurfaces. Many ions such as F^-, Cl^-, Si (as SiO_4^{4-}) and Zn^{2+} are found at higher concentrations near the surface, whereas other ions such as CO_3^{2-}, Na^+, and Mg^{2+} increase toward the deeper parts of the enamel.[54,74] A few ions such as K^+ appear to be rather evenly distributed. These gradients can be due to pre- or posteruptive changes. For example, it seems that the F^- gradient is largely due to posteruptive uptake of F^- from oral fluid,[68] whereas the Cl^- gradient is predominantly determined pre-eruptively.[61] With respect to the decrease in acid solubility of dental enamel by F^-, it should be mentioned that it is generally accepted that F^- stabilizes the lattice by disordering the central OH^- ions in the apatite and preventing the lining up of OH^- ions in the same direction, and by the formation of OH-F^- bonds.

On the other hand, the experiments of Nikiforuk and Grainger[45] show that F^- reduces the solubility by replacing CO_3^{2-} in the lattice of the crystallites enhancing thereby the crystallinity of apatite. The most abundant trace element in the dental enamel is Mg^{2+} (0.4%). Although it has been proven by Neuman and Mulryan[43] that Mg^{2+} ions cannot be incorporated in the apatite structure, the idea cannot be overlooked that Mg^{2+} can exist in another calcium phosphate salt, whitlockite ($Ca_9Mg(HPO_4)(PO_4)_6$ (mol Ca/P ratio = 1.286), which will be formed in the presence of Mg^{2+}.[28] This salt cannot be detected by X-ray diffraction because it is too finely dispersed. Other authors[69] believe that Mg^{2+} has nothing to do with the inorganic part of the enamel, but is more related to the enamel matrix.

VI. CARIES

Until recently, no satisfying explanation has been given for the great individual variation in resistance of teeth against the acid attack of microorganisms in the caries process. In the foregoing chapter it has been shown that endogenous factors must play a very important role, especially during the development of the teeth. The purpose of this section is to describe some developmental aspects of the caries resistance both pre-eruptively as well as posteruptively. Moreover, special attention will be paid to some processes occurring normally during the posteruptive maturation, because these maturation processes can be used for the repair of partially demineralized enamel. However, before these enamel-related factors are described, some general features of the caries process will be given.

Microorganisms in the mouth are very important for the caries development because they can produce plaque which keeps acid products of the metabolism of carbohydrates in the diet in close contact to the enamel. This leads to the dissolution of the dental enamel. From the carbohydrates, the sucrose most often used in the diet will lead to the highest acid production. Finally, it has been shown that a certain amount of time is needed for decreasing the pH below pH 5.5 at which dissolution of the enamel occurs. Depending on the kind and concentration of carbohydrates, it varies between 5 to 90 min. In order to protect the dental enamel against its dissolution, the saliva is an absolute prerequisite. Both its organic, as well as its inorganic components contain factors that can increase the resistance of dental enamel against caries.

Before describing the behavior of the dental enamel during the caries process, two other very important factors will be described: the saliva and the plaque.

A. The Saliva: Inorganic Composition

Calcium and phosphate concentrations in the saliva are supersaturated with respect to the apatite crystals of dental enamel. Glycoproteins, Mg^{2+} and HCO_3^- or pyrophosphate are the most important components of the saliva for keeping calcium and phosphate in a metastable equilibrium. The high calcium and phosphate concentrations are very important during the posteruptive development for increasing the caries resistance by the maturation of the young enamel.[73] Apart from the Ca^{2+} and P_i content, the buffer capacity of the saliva is very important. At high pH (7.2 to 7.4) the highest contribution to it will be from the CO_2/HCO_3^- system, whereas at low pH (pH 5.5) at which the enamel dissolves, the $HPO_4^{2-}/H_2PO_4^-$ systems are the most responsible for the buffer capacity. Apart from the inorganic components of the saliva, organic constituents can also play an important role at the caries process.

First, it is well-known that soon after the eruption of the tooth, a pellicle is found as the consequence of protein sedimentation on the enamel which regulates ionic transport and facilitates aggregation of the bacterial plaque. Second, three types of antibacterial factors can be described in the saliva:

1. Lysosomes (acting on the cell membrane, sometimes together with the immunoglobulin IgA).[50] This enzyme hydrolyses the 1.4 bound between *N*-acetyl glucosamine and *N*-acetyl muramic acid both belonging to the polysaccharides of the cell membrane of the bacteria. In this way, the enzyme can behave as bacteriostatic or cariostatic. The effect can be increased by the presence of detergentia such as laurylsulfate.
2. The peroxidase system.[8] $H_2O_2 + AH_2 \rightleftharpoons 2H_2O + A$ (A = hydrogendonor, PH_2 = hydrogen deceptor). It acts against the microorganisms accumulating H_2O_2 (*Lactobacillus acidophilus, Streptococcus cremoris*). Through this action, bacteria are not able to take up essential amino acids. CNS^- stabilizes the peroxidase system. In this way the *S. mutans* can be inhibited.
3. Immunoglobulins IgA, IgG, and IgH.[25] Among the immunoglobulins in saliva, IgA is the most important. Because it contains an extra glycopeptide, the IgA differs from the analogue globulin in the serum. It is produced by the immunocytes in the salivary glands. By decreasing adsorption of streptococci or by increasing phagocytosis it inhibits the growth of the bacteria. IgG and IgM originate partially from the serum and are also related to the caries process. The general action of IgA is to inhibit the adsorption of the bacteria on the mucous membranes in the mouth. Among the immunological control against dental caries there exist two hypotheses. One hypothesis proposed by English workers suggests that serum IgG antibodies are responsible for inhibiting caries. These pass from the plasma to the oral cavity via the crevicular fluid in order to contribute to the local defense mechanisms by inhibiting the colonies of bacteria, especially in the fissures. Another hypothesis and put forward primarily by American scientists, stresses the importance of local immunity due to secretory IgA antibodies. This function is to inhibit the adherence of *S. mutans* to the enamel surfaces.[25] This immunity could be obtained by using whole or broken cells as well as cell free glycosyltransferase (G-T-ase) for the local immunization. Reduction in caries was greater on smooth surfaces of teeth than on occlusal surfaces, probably due to interference with adherence of *S. mutans*. It is possible to increase the IgA immune response only in the saliva, while eliciting a serum antibody response. The problem of cross-reaction between heart tissue and certain components of *S. mutans* strains for IgG in the vaccine attack can be avoided by using G-T-ase for immunization. In such a case, the quantity of antibody necessary to reduce the caries may not be detrimental to the host.

B. Dental Plaque

After tooth eruption, as well as after cleaning of the dental enamel, glycoproteins from the saliva adhere to the enamel and change in composition to the so-called acquired pellicle. This pellicle binds the bacteriae, which multiply, grow, produce the extracellular material, and form the plaque. Glycoproteins are composed of a carbohydrate and a protein component. By the presence of the terminal *N*-acetyl muramic acid, the most saliva glycoproteins are negatively charged. The adherence of these proteins to the enamel surfaces can be explained by the amphoteric behavior to which negative as well as positive macromolecules can be bound. Depending on the charge distribution between the enamel surface and the glycoprotein molecule, the specificity of the adsorption can then be explained. Ca^{2+} can improve the strong binding of glycoproteins to the enamel. By enzymatic splitting the hydrophylic carbohydrate part of the glycoproteins, muramic acid disappears in the pellicle thereby decreasing the solubility of the rest of the pellicle. This insoluble mature pellicle forms the basis for bacterial adherence.[24]

The growth of the microorganisms in the mouth is determined by local oxygen concentration, pH, and growth-stimulating or growth-inhibiting factors. Selective adsorption mainly determines the place of the microorganisms. This again depends, as previously mentioned, on the charge of surfaces of the enamel or bacteria. The adsorption is determined by London-v.d. Waalse forces or by Ca^{2+} bridges. Among the first microorganisms adhering to the acquired pellicle are *Streptococcus sanguis* which agglutinates in the presence of glycoproteins, and *Actinomyces viscosis* which produces lactic acid used by the *Veillonella*. The microorganisms produce, mainly from sucrose, extracellular polysaccharides (EPS). Although a (1-6) bound soluble glycopolymer or dextran is built up by *Streptococcus sanguis* and *S. mutans*, the insoluble 1-3 bound glycose polymer is merely synthesized by *S. mutans*. It is called mutan. Apart from the glucan formation several streptococci and *Actinomyces viscosis* can polymerize fructose toward the soluble fructan, called levan. Both glucose and fructose polymers form the matrix in which during one or two days a strong growth of bacteria takes place.

The environment on the enamel surface becomes more and more anaerobic, because of the increasing impairments of oxygen diffusion in the plaque. After 5 to 6 days the plaque has obtained its full form and an equilibrium exists between the physiological cleaning and further growth of the plaque. This plaque contains 3×10^8 bacteria per milligram. Apart from these the plaque contains glycoproteins and EPS called matrix. Most of the bacteria are streptococci and actimomycetes. Lactobacilli are present in a very low concentration. *Streptococcus mutans* also not present that abundantly in the plaque, can strongly change and play a role in the caries process because of their acid production. Carbohydrates are used in the dental plaque as an energy source for the microorganisms. Very important in this respect are the soluble and often easily fermentable carbohydrates such as glucose, fructose, maltose, saccharose, and lactose. They will be fermented anaerobically to acids or other residues such as CO_2 or alcohol depending on the carbohydrate supply.[25] The used energy-rich carbohydrates can lead to the formation of intra- or extracellular polysaccharides (IPS or EPS). This energy will be produced by fermentation of carbohydrates to acids. The 1-4 glucose polymer glycogen is among the intracellular polysaccharides. The glycogen will be formed during a high supply of carbohydrates to the plaque and can be a factor in the caries process by keeping the acid production high in the plaque over a longer time. The extracellular polysaccharides, which are formed by bacteria, can also be fermented, for example, by the enzyme dextranase. Only the glucose polymer mutan cannot be broken down by bacteria.

The acid production of the dental plaque can be eliminated by several processes:

1. Changing the strong acidic lactic acid toward the lower acidic propionic acid.
2. Urea metabolism toward NH_4^+ and CO_2.
3. Processes of decarboxylation.

There are several arguments to which to pay special attention among the strong acid producers toward the *Streptococcus mutans:*

1. There is a close connection between carious surfaces and the presence of the microorganisms.
2. There is a high correlation between the frequency of *S. mutans* and the development of caries in the same surfaces.
3. *S. mutans* is very acid-resistant and an acid producer, as well as a glycogen former.
4. In animals *S. mutans* can form very insoluble mutans from sucrose thereby increasing the frequency of caries.[44]

As previously mentioned, dental caries is the consequence of fermentation of carbohydrates to acids. These acids that are surrounded by the plaque will not be buffered by the saliva, but will accumulate toward a high concentration on the enamel surface by the plaque and will demineralize the dental enamel. Apart from the composition of the saliva mentioned above, the intake of carbohydrates, the formation of plaque and acid, and the processes of de- and remineralization are very important for explaining the caries process.

C. Dental Enamel

As the author mentioned in another section, the dental enamel changes during its pre- and posteruptive maturation and becomes more resistant toward caries. These processes are mainly the consequences of diffusion. However, unexpectedly, the diffusion channels in the dental enamel do not plug and enable the calcium and phosphate ions to penetrate into the deepest situated hypomineralized areas during the whole processes of maturation.[10] An explanation for this may be found in the work of Wöltgens et al.[73] who state that during posteruptive maturation no real accumulation of calcium and phosphate takes place, but rather a loss of both ions in the subsurface region. This suggests that apart from diffusion recrystallization and consequently a high exchange of the ions also occurs.

Processes of remineralization of enamel partially demineralized by the caries process are comparable with posteruptive maturation.[19] However, in those experiments it was shown that ^{32}P uptake was higher in abraded enamel than in intact enamel. That is why in contrast to posteruptive maturation, plugging of the surface of dental enamel also has to be taken into account. Moreover, experiments of Koulourides and Cameron[32] show that even a higher tooth resistance can be obtained by remineralization of the demineralized enamel than by the sound enamel itself. They consider the caries process as a result of many intermittent cariogenic and reparative periods. The pH of the cariogenic plaque decreases immediately following the ingestion of fermentable carbohydrates.[65] At a pH range below 5.5, the plaque fluid becomes undersaturated with respect to apatite of enamel, hence fostering demineralization. During the subsequent neutralization of acid, the plaque fluid becomes supersaturated with respect to hydroxyapatite and the dissolved minerals of the plaque fluid tend to precipitate either in the micro spaces of partially demineralized enamel or within the plaque. This interaction of enamel with the demineralizing and remineralizing activities of the plaque influences enamel composition, as indicated by the changes in the nitrogen, fluoride, and increased resistance to acid of the ''altered'' enamel.

1. Demineralization

Demineralization such as that which occurs during the caries process is not comparable with etching, but the acid penetrates to the subsurface enamel, leaving the surface relatively intact (Figure 6). This leads to an incipient caries lesion, the so-called white spot. In the body of the lesion demineralization of the interprismatic area takes place initially and later on demineralization of the cross-striations in the prisms along the prisms' cores.[16] It seems that in the above areas, calcium phosphate salts exist that are more soluble than the remaining salts. In this respect it is interesting that Wöltgens et al.[72] were able to detect, apart from apatite with a mol Ca/P ratio of 1.63, another salt with a mol Ca/P = 1.85. This salt is situated parallel to the surface and comprises ⅓ of the total calcium salts in human dental enamel. These findings suggest that this salt is probably a carbonate-containing apatite which is situated in the cross-striations and more soluble than normal apatite. The relative intact surface in the incipient caries lesion has been ascribed toward specific properties different from the more interior-situated dental enamel layers,[29,73] whereas Moreno and Zahradnik[40] suggest that the relative intact surface of enamel is the result of reprecipitation of salts dissolved in the body of the lesion during the formation of the brushite $CaHPO_4 2H_2O$.[3] Processes of demineralization are reversible as long as the organic material in the intraprismatic spaces remain intact. If they have been changed or eliminated during the process of demineralization, no real remineralization can take place.[41]

2. Remineralization

Studies on remineralization of etched or otherwise demineralized dental enamel have been carried out with saliva or with only calcium and phosphate concentrations present normally in the saliva.[41] Remineralization is proportional to the degree of supersaturation of the calcium and phosphate ions in the remineralization solution with respect to the demineralized enamel. This solution can be saliva or resting plaque. However, if the degree of supersaturation is high, the remineralization is restricted to the surface because of plugging the channels, thereby limiting diffusion of the ions to more interior areas. In this respect, it is interesting to note that the behavior of saliva is comparable with a lower saturated solution than solutions with the same calcium and phosphate concentrations alone. Apparently proteins in the saliva can inhibit remineralization.[30] McGaughy and Stowell[22] were able to determine selectively calcium-binding glycoproteins in the saliva which can locally be absorbed towards the apatite, and form a cuticle or pellicle on the enamel surface. This cuticle is only permeable for small ions[58] and makes the enamel more resistant to etching by acids. Meckel[38] describes three types of cuticles on a clinical and an electron microscopical basis:

1. Surface cuticle 0.2 μm thick
2. Stained cuticle 10 μm thick, brown
3. Subsurface cuticle 1 to 3 μm thick deposited in attacked enamel

The surface and subsurface cuticles are formed within 10 days, whereas the stained cuticle could only be determined after 215 days. Saxton[57] found that they can fill up scratches within 20 min in the enamel surface produced by prophylaxis.

A tyrosine-rich peptide is isolated from the saliva of the parotic gland and is called steatherin. It possesses a very high affinity for apatite and is able to reduce mineralization greatly in a 0.4 μmol concentration,[41] whereas the normal concentration in the saliva is 5 μmol. It has been suggested that it adsorbs preferentially to nucleating sites of the apatite crystals.

Hexametaphosphates, pyrophosphates,[1] and Ca^{2+} [31] seem to act in a similar manner. On the other hand, the F^- present in the saliva can stimulate mineralization. Pigman

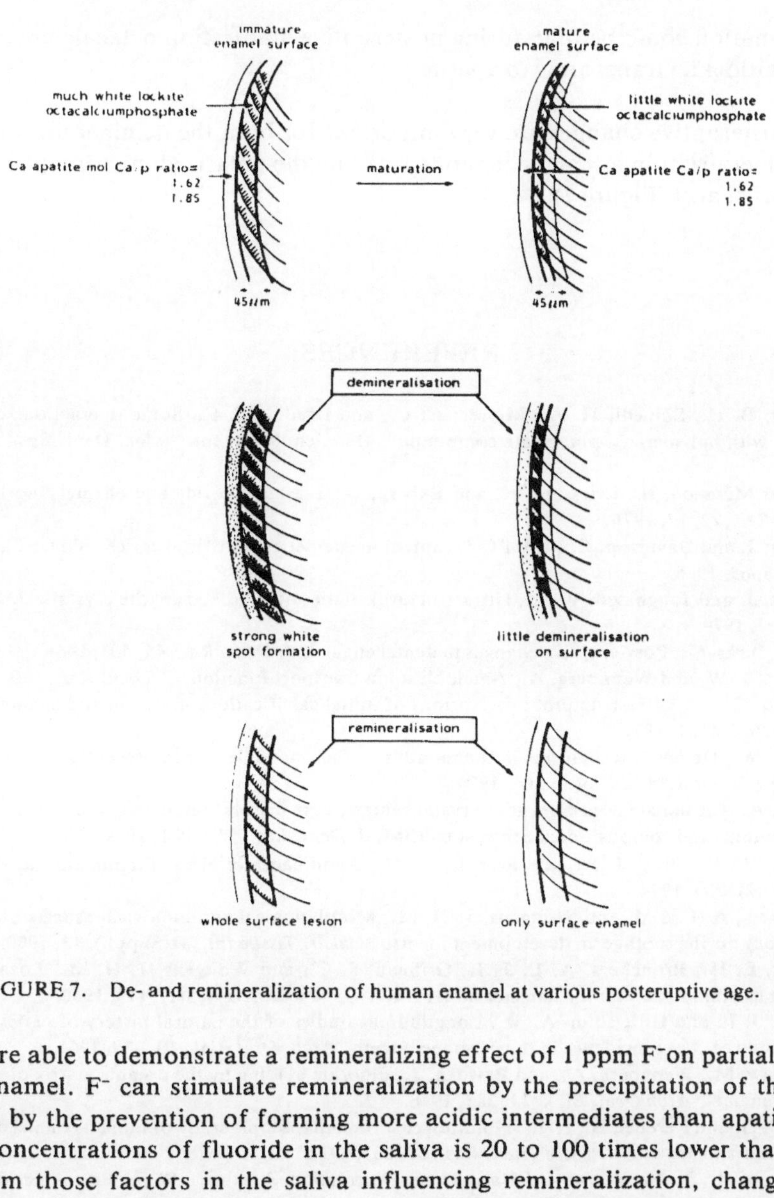

FIGURE 7. De- and remineralization of human enamel at various posteruptive age.

et al.[49] were able to demonstrate a remineralizing effect of 1 ppm F^- on partially demineralized enamel. F^- can stimulate remineralization by the precipitation of the fluorapatite and by the prevention of forming more acidic intermediates than apatite. However the concentrations of fluoride in the saliva is 20 to 100 times lower than 1 ppm. Apart from those factors in the saliva influencing remineralization, changes in the dental enamel during aging can modify its reactivity toward remineralization solution or saliva.

1. The formation of the so-called discolored areas in the enamel surface. They contain more H_2O and organic compounds and less inorganic material.[56] Consequently, the density of the discolored areas is lower than the surrounding normal enamel. They seem, however, to be more resistant to acid dissolution.[52] This can be explained by the lower Na^+ or higher F^- concentrations inside the areas.
2. The penetration of remineralizing solution can be changed from 100 μm in non-erupted tooth enamel toward 20 μm after eruption.[59] Possibly the formation of the cuticle after eruption can explain these findings.
3. During posteruptive maturation, the calcium and phosphate content in the enamel decreases, probably by the formation of discolored areas. However, the mol Ca/P ratio also changed from 1.5 originally to 1.7 after maturation.[73] An

explanation could be that during posteruptive maturation octacalcium phosphate or whitlockite transforms to apatite.

These posteruptive changes are very important for both the demineralization as well as the remineralization as can be demonstrated by the polarized microscopical findings of Wöltgens et al.[71] (Figure 7).

REFERENCES

1. **Amdur, B. H., Spinelli, H. A., Messer, A. C., and Brudevold, F.,** Some interactions of hydroxy apatite with polymerized phosphate compounds, 41st Gen. Meet. Int. Assoc. Dent. Res., Abst. 331, 1963.
2. **Angmar-Månsson, B., Ericsson, Y., and Ekberg, O.,** Plasma fluoride and enamel fluoride, *Calcif. Tissue Res.*, 22, 77, 1976.
3. **Arends, J. and Davidson, C. L.,** HPO_4^{2-} content in enamel and artificial caries lesions, *Calcif. Tissue Res.*, 18, 65, 1975.
4. **Arends, J. and Jongbloed, W. L.,** Ultrastructural studies of synthetic apatite crystals, *J. Dent. Res.*, 58B, 837, 1979.
5. **Backer Dirks, O.,** Post-eruptive changes in dental enamel, *J. Dent. Res.*, 45, 503, 1966.
6. **Bawden, J. W. and Wennberg, A.,** Ameloblast ion transport function, *J. Dent. Res.*, 58B, 708, 1979.
7. **Bernard, G. W.,** Ultrastructural observations of initial calcifications in dentin and enamel, *J. Ultrastruct. Res.*, 41, 1, 1972.
8. **Bowen, W.,** Defense mechanisms in the mouth and their possible role in prevention of dental caries. A review, *J. Oral Pathol.*, 3(6), 266, 1974.
9. **Boyde, A.,** Carbonate concentration, crystal centers, core dissolution, caries, cross striations, circadian rhythms and compositional contrast in SEM, *J. Dent. Res.*, 58B, 981, 1979.
10. **Briner, W. W., Gray, J. A., and Francis, M. D.,** Significance of enamel remineralization, *J. Dent. Res.*, 53(2), 239, 1974.
11. **Bronckers, A. L. J. J. and Wöltgens, J. H. M.,** Morphological and biological aspects of vitamin C deficiency during toothgerm development in vitro, *Calcif. Tissue Int.*, 31(Suppl.), 13, 1980.
12. **Burger, E. H., Bronckers, A. L. J. J., Gribnau, J. C., and Wöltgens, J. H. M.,** Localization of cellular calcium in developing hamster molar teeth, *Ultramicroscopy*, 5(1), 113, 1980.
13. **Carlos, J. P. and Gittelsohn, A. M.,** Longitudinal studies of the natural history of caries. II. A life table study of caries incidence in the permanent teeth, *Arch. Oral Biol.*, 10, 739, 1965.
14. **Crenshaw, M., Wennberg, A., and Bawden, J.,** Fluoride binding by the organic matrix of developing bovine enamel, *Arch Oral. Biol.*, 23, 285, 1978.
15. **Daculsi, G. and Kerebel, B.,** High resolution electron microscopic study of human enamel crystallites: size, shape and growth, *J. Ultrastruct. Res.*, 65, 163, 1978.
16. **Darling, A. I.,** in *Mechanisms of Hard Tissue Destruction*, Sognnaes, R. F., Ed., American Association for the Advancement of Science, Washington, D.C., 1963, 171.
17. **Driessens, F. C. M.,** The mineral in bone, dentin and tooth enamel, *Bull. Soc. Chim. Belg.*, 89(8), 663, 1980.
18. **Eastoe, J. E.,** The amino acid composition of proteins from the oral tissues, *Arch. Oral Biol.*, 8, 633, 1963.
19. **Von der Fehr, F.,** in *Fluorine Research and Dental Caries Prevention*, Vol. 3, Hardwick, J. L., Held, R. R., and König, K. G., Eds., Pergamon Press, Oxford, 1964, 83.
20. **Fejerskov, O., Johnson, N. W., and Silverstone, L. M.,** The ultrastructure of fluorosed human dental enamel, *Scand. J. Dent. Res.*, 82(5), 357, 1974.
21. **Frank, R. M.,** Tooth enamel: current state of the art, *J. Dent. Res.*, 58B, 684, 1979.
22. **McGaughey, C. and Stowell, E. C.,** The simultaneous binding of salivary protein, calcium and phosphate by hydroxy apatite, 49th Gen. Meet. Int. Assoc. Dent. Res., Abstr. 653, 1971.
23. **Gaunt, W. A., Osborn, J. W., and ten Cate, A. R.,** *Advanced Dental Histology*, 3rd ed., J. Wright and Sons, Bristol, England, 1976.
24. **Gibbons, R. J. and van Houte, J.,** On the formation of dental plaque, *J. Periodontol.*, 44, 346, 1973.
25. **Hamada, S. and Slade, H. D.,** Biology, immunology and cariogenity of *S. mutans* and other oral streptococci, *J. Dent. Res.*, 53, 22, 1979.

26. **Harris, R. S. and Navia, J. M.**, Vitamin A deficiency and caries susceptibility of rat molars, *Arch. Oral Biol.*, 25, 415, 1980.
27. **Harrison, J. E. and Harrison, H. C.**, in *Vitamin D: Perspective of Four Decades of Research in Pediatric Diseases Related to Calcium*, de Luca, H. F. and Anast, C. S., Eds., Elsevier, New York, 1980, 1.
28. **Hayek, J. and Newesly, H.**, Ueber die Existenz von Tricalciumphosphat in Wässiger Lösung, *Monatsh. Chem.*, 89, 88, 1958.
29. **Holly, F. J. and Gray, J. A.**, Mechanism for incipient caries lesion growth utilizing a physical model based on diffusion concept, *Arch. Oral Biol.*, 13, 319, 1968.
30. **Koulourides, T.**, Dynamics of tooth surface oral fluid equilibrium, *Adv. Oral Biol.*, 2, 149, 1966.
31. **Koulourides, T., Feagin, F., and Pigman, W.**, Effect of pH, ionic strength and cupric ions on the rehardening rate of buffered softened human enamel, *Arch. Oral Biol.*, 13, 335, 1968.
32. **Koulourides, T. and Cameron, B.**, Enamel remineralization as a factor in the pathogenesis of dental caries, *J. Oral Pathol.*, 9, 255, 1980.
33. **Kruger, B. J.**, Ultrastructural changes in ameloblasts from fluoride treated rats, *Arch. Oral Biol.*, 13, 969, 1968.
34. **Kruger, B. J.**, The effects of different levels of fluoride on the ultrastructure of ameloblasts in the rat, *Arch. Oral Biol.*, 15, 109, 1970.
35. **Levenson, G. E.**, Effect of ascorbic acid deficiency on mouse second molar tooth germ cultivated in vitro, *J. Embryol. Exp. Morphol.*, 36(1), 73, 1976.
36. **Likins, R. C., Posner, A. S., and Steere, A. C.**, Effect of calcium treatment on solubility and Ca uptake of synthetic hydroxyapatite and rat molar enamel, *J. Am. Dent. Assoc.*, 57, 335, 1958.
37. **Lyaruu, D. M., Belcourt, A., Fincham, A. G., and Termine, J. D.**, Neonatal hamster molar tooth development: extraction and characterization of amelogenesis and soluble dentin protein, *Calcif. Tissue Int.*, 34, 86, 1982.
38. **Meckel, H.**, The formation and properties of organic films on teeth, *Arch. Oral Biol.*, 19, 585, 1965.
39. **Mønstad, H. and Hammarström, L.**, Morphological changes in rat enamel organ following a single intraperitoneal injection of sodium fluoride, *Scand. J. Dent. Res.*, 86, 211, 1978.
40. **Moreno, E. C. and Zahradnik, R. T.**, Chemistry of enamel subsurface demineralization in vitro, *J. Dent. Res.*, 53, 226, 1974.
41. **Moreno, E. C. and Zahradnik, R. T.**, De- and remineralization of dental enamel, in Proc. 3rd Int. Symp. Tooth Enamel, Nylen, M. U. and Termine, J. D., Eds., *J. Dent. Res.*, 58(special issue), 896, 1979.
42. **Navia, J. M.**, Prevention of dental caries: agents which increase tooth resistance to dental caries, *Int. Dent. J.*, 22, 427, 1972.
43. **Neuman, W. F. and Mulryan, B. J.**, Synthetic hydroxy apatite crystals. IV. Mg incorporation, *Calcif. Tissue Res.*, 7, 133, 1971.
44. **Newburn, E.**, *Cariology*, Williams & Wilkins, Baltimore, 1978, 57.
45. **Nikiforuk, G. and Grainger, R. M.**, Fluoride carbonate citrate interactions in enamel, in *Tooth Enamel*, Stack, M. V. and Fearnhead, R. W., Eds., J. Wright and Sons, Bristol, 1965, 26.
46. **Nikiforuk, G. and Fraser, D.**, Etiology of enamel hypoplasia and interglobular dentin: the roles of hypocalcaemia and hypophosphatemia, *Metab. Bone Dis. Rel. Res.*, 2, 17, 1979.
47. **Patterson, C. M., Basford, K. E., and Kruger, B. J.**, The effect of fluoride on the immature enamel matrix proteins of the rat, *Arch. Oral Biol.*, 21, 131, 1976.
48. **Paynter, K. J. and Grainger, R. M.**, Relation of nutrition to the morphology and size of rat molar teeth, *J. Can. Dent. Assoc.*, 22, 519, 1956.
49. **Pigman, W., Cueto, H., and Baugh, D.**, Conditions affecting the rehardening of softened enamel, *J. Dent. Res.*, 43, 1187, 1964.
50. **Pollock, J. J., Goodman Bicker, G., Katona, L. I., Cho, M. I., and Iacono, V. J.**, in *Saliva and Dental Caries*, Kleinberg, Ellison, and Hansel, Eds., Interscience, New York, 1979, 429.
51. **de Porter, D. A. and ten Cate, A. R.**, Fine structural localization of alkaline phosphatase in relation of enamel formation, *Arch. Oral Biol.*, 21, 7, 1976.
52. **Posen, J. M.**, Chemical and physical properties of altered and sound enamel. II. Relative dissolution and residue light adsorption (color), *J. Dent. Res.*, 41, 471, 1962.
53. **Robison, R.**, The possible significance of hexose phosphoric esters in ossifications, *Biochem. J.*, 24, 1927, 1931.
54. **Robinson, C., Briggs, H. D., Atkinson, P. J., and Weatherell, J. A.**, Matrix and mineral changes in developing enamel, *J. Dent. Res.*, 58B, 871, 1979.
55. **Salley, J. J., Bryson, W. F., and Eshleman, J. R.**, The effect of chronic vitamin A deficiency on dental caries in Syrian hamster, *J. Dent. Res.*, 38, 1038, 1959.
56. **Savory, A. and Brudevold, F.**, Distribution of nitrogen in human enamel, *J. Dent. Res.*, 38, 436, 1959.

57. Saxton, C. A., Scanning electron microscope study of plaque formation, *Caries Res.*, 6, 75, 1972.
58. Schüle, H., *Das Schmelz Oberhautchen*, Georg Thieme Verlag, Stuttgart, 1962.
59. Silverstone, L. M. and Johnstone, N. W., The effect on sound human enamel of exposure to calcifying fluids in vitro, *Caries Res.*, 5, 323, 1971.
60. Sobel, A. E. and Hanok, A., Calcification. XVI. Composition of bones and teeth in relation to blood and diet in the cotton rat, *J. Dent. Res.*, 37, 631, 1958.
61. Söremark, R. and Grøn, P., Chloride distribution in human dental enamel as determined by electron microanalysis, *Arch. Oral. Biol.*, 11, 861, 1966.
62. Speirs, R. L., Factors influencing maturation of developmental hypomineralized areas in the enamel of rat molars, *Caries Res.*, 1, 15, 1967.
63. Speirs, R. L., The nature of surface enamel in human teeth, *Calcif. Tissue Res.*, 8, 1, 1971.
64. Speirs, R. L., Fluoride concentrations in toothgerms of permanent teeth of the domestic pig, *Arch. Oral Biol.*, 23, 1019, 1978.
65. Stephan, R. M., Changes in hydrogen ion concentration on tooth surfaces in carious lesions, *J. Am. Dent. Assoc.*, 27, 718, 1940.
66. Tanzer, F. S. and Navia, J. M., Pre-eruptive cariostatic effect of thyrocalcitonin on rat molars, *Int. Assoc. Dent. Res.*, Abstr. 505, 1972.
67. Wassermann, F., Analysis of enamel formation, *J. Dent. Res.*, 23, 463, 1944.
68. Weatherell, J. A. and Robinson, C., in *Biological Mineralization*, Zipkin, I., Ed., John Wiley & Sons, New York, 1973, 43.
69. Weatherell, J. A., Deutsch, O., Robinson, C., and Hallsworth, A. S., Assimilations of fluoride by enamel throughout the life of the tooth, in Cariostatic Mechanism of Fluoride, König, K. G., *Caries Res.*, 11(Suppl. 1), 85, 1977.
70. Wöltgens, J. H. M., Role of Inorganic Pyrophosphatase in Dental Mineralization, Thesis, University of Nijmegen, Netherlands, 1971.
71. Wöltgens, J. H. M., Houwink, B., and Bervoets, T. J. M., Influence of the diphosphonate EHDP on the remineralization of artificial caries in human enamel, *Caries Res.*, 13, 163, 1979.
72. Wöltgens, J. H. M., Vingerling, P. A., and Witjes, F., Chemical evidence of two separate apatite phases in human enamel, *Arch. Oral Biol.*, 25, 435, 1980.
73. Wöltgens, J. H. M., Bervoets, T. J. M., Witjes, F., and Driessens, F. C. M., Changes in the Ca and P content of the enamel of human premolars shortly after eruption, *Arch. Oral Biol.*, 26, 717, 1981.
74. Zipkin, I., in *Biological Calcification*, Schraer, H., Ed., Appleton-Century-Crofts, New York, 1970, 69.

Chapter 3

HISTOLOGY OF BONE: A SYNOPSIS

P. J. van Mullem and J. C. Maltha

TABLE OF CONTENTS

I. GENERAL INTRODUCTION

The cells of the animal body possess several structures common to all. However, the relative amount of these structures is different for the various types of cells. Furthermore, a certain type of cell might have specific structures which are related to its function in the body. This specialization is brought about by the process called (cell)differentiation which not only takes place in the embryo, but can even take place in certain tissues during adult life. A pool of undifferentiated cells remains in these tissues during the whole life span of the animal.

The multitude of functions of cells in the body is reflected by a multitude of internal and external structures of the cells. As the cells of the body are all derived from one cell, the fertilized germ cell, and after each cell division (mitosis), both daughter cells normally possess a complete set of chromosomes, of which the genetic material is DNA, some explanation must exist for the differences in structure and function. It is generally accepted that some parts of the DNA in a given differentiated cell are repressed, while others are not. These parts of the genetic material attain expression, i.e., govern the structure and function of the differentiated cell.

Although a multitude of cell types exists in the animal and human body, similar cells and those closely related, are located in groups (the tissues). Four basic tissues are usually distinguished:

1. **Epithelium:** This tissue consists of closely aggregated cells and has very little intercellular substance. It covers external and internal surfaces.
2. **Connective tissue:** It consists of cells and a vast amount of intercellular substance which is subdivided into fibers and ground substance. Fibers and ground substance impart firmness to this tissue and to the organs of which they are part. Areolar tissue, or loose connective tissue, tendons, cartilage, bone, dentin, and cementum belong to it. The three last mentioned forms of connective tissue mainly differ from the others in that they are mineralized. The function of bone is four-fold:
   It supports and protects organs, mainly in the skull and the thorax
   It provides firm sites for the attachment of muscles
   It serves part of the immediate need of the body for calcium
   It contains the bone marrow in its cavities, which serves the supply of blood cells (red bone marrow) and the storage of fat (yellow bone marrow)
3. **Muscular tissue:** The cells of this tissue are contractile and usually arranged in bundles. The voluntary, so-called striated, muscles, the nonvoluntary, so-called smooth muscles, and the cardiac muscles belong to it.
4. **Nervous tissue:** The shape of the nerve cell (neuron) is usually very complex. The cell consists of a cell body (the perikaryon), which lodges the nucleus and several processes. The volume of the processes is often larger than that of the perikaryon.

This chapter deals with the histology and the cytology of that kind of the mineralized connective tissues which is called bone. After a review of the cell types in and around the bone matrix, bone deposition and resorption, the types of bone, growth and remodeling, and the vascularization of bones will be discussed, mainly on a light-microscopical level only.

II. BONE CELLS, PERIOSTEUM, AND ENDOSTEUM

A. Introduction

Bone is produced by cells, it contains cells, and it is resorbed by cells. The morphology of bone is a result of some sort of a balance between the numbers of cells and the various functional roles. In areas where bone is deposited, for example, during growth or fracture healing, the relative amount of bone forming cells, the osteoblasts, is high. In other areas, massive bone resorption can take place by bone resorbing cells, the osteoclasts. In the adult, a steady state is attained: a continuous remodeling, i.e., a combined action of resorption and deposition is found under normal conditions of load and tension exerted on the bone. When these conditions change, bone is able to adapt to the altered loads and tensions by remodeling processes or it atrophies or hypertrophies. This plastic character of bone is not in contradiction with its supportive, protective, and attachment functions. Remodeling, together with the growth and development of bones, is caused by the activity of the bone cells, the periosteum which covers all bones, and the endosteum which lines the bones internally.

Regarding the genesis of the bone cells, two theories exist. In both theories, a cell of the pool of the undifferentiated mesenchymal cells can change into an osteoprogenitor cell and in doing so enters the realm of the cells in bone. It then can differentiate into an osteoblast which can then become an osteocyte. In the case of dedifferentiation, an osteocyte can become an osteoblast again or can dedifferentiate into an osteoprogenitor cell. In the one theory, an osteoclast is a product of differentiation from osteoblasts and fusion of a number of these transformed cells. Dedifferentiation can take place into (separate) osteoblasts.

In the other theory, the osteoclasts originate either directly from the osteoprogenitor cells or from monocytes of the blood. In the latter case, monocytes leave the blood vessel by diapedesis and differentiate to mononuclear phagocytes. By fusion of several of these cells an osteoclast originates. Upon dedifferentiation, inactive mononuclear cells arise.

These cells, the bone matrix, and the tissue layers will be discussed in the following sections.

B. Osteoblasts

Osteoblasts are the cells forming the organic bone matrix during growth, remodeling, and repair processes. Usually they are arranged in an epitheloideal fashion, i.e., as a continuous layer of cells which lines the bone tissue and separates it from the remaining tissue (Figure 1). In some areas this layer may be interrupted. In general, the cells of a layer of osteoblasts can demonstrate two forms and this relates to their function. A cell can be active or inactive. When an osteoblast is active (Figure 2), its shape is cuboidal, sometimes low cylindrical, and its cytoplasm stains well with basic dyes or hematoxylin because of the ample presence of RNA. The RNA is present in the rough endoplasmatic reticulum and serves the synthesis of proteins. Near the nucleus, the pale staining Golgi apparatus is found. The nucleus is large and usually contains a distinct nucleolus. Thin projections of these cells are in contact with projections from cells embedded in the bone matrix and also with the neighboring osteoblasts. In the instance that the osteoblast is inactive (Figure 3), it has a flattened appearance, the nucleus is smaller, and the cytoplasm does not contain a large amount of RNA.

C. Osteocytes

The deposition of organic matrix by osteoblasts is not only found between the osteoblasts and the already existing bone, but also between the osteoblasts themselves. By this process osteoblasts may become embedded in the matrix. Once surrounded by bone matrix, the cells are called osteocytes. Thus they are located in cavities (lacunae) in the bone (Figure 4). The cells and the lacunae in which the cells fit snugly, are flattened, and are usually larger in young bone than in old bone. The cytoplasm stains

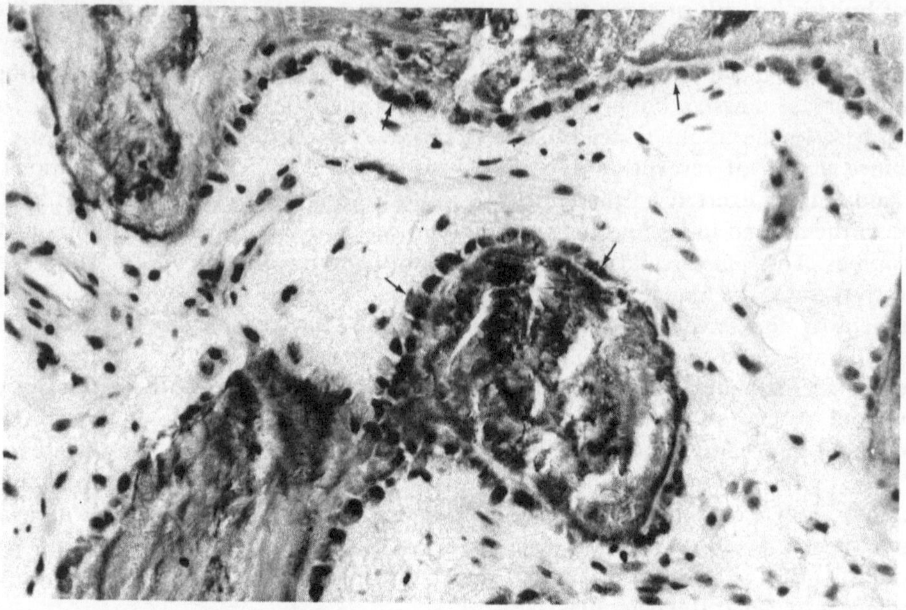

FIGURE 1. Continuous layer of osteoblasts (arrows) against bone. (Hematoxylin-toluidine blue-acid fuchsin stain; magnification × 200.)

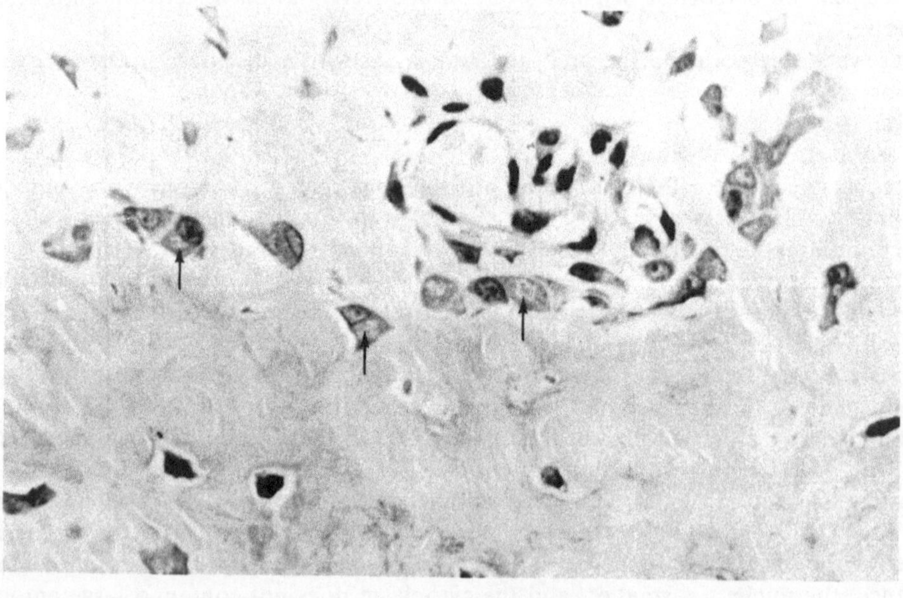

FIGURE 2. Active osteoblasts lining the bone. In some cells the pale staining Golgi apparatus can be distinguished (arrows). 1 μm thick section. (Toluidine blue stain; magnification × 500.)

faintly, whereas the nucleus, which occupies a large part of the cell body, usually stains intensely. Characteristic for this cell type are the numerous processes which are present in tunnels (canaliculi) in the bone (Figure 5). Those canaliculi which belong to neighboring osteocytes are connected with one another and can transfer oxygen and nourishment from and metabolites to the blood. As the mineralized part of a bone is avascular and diffusion of these substances through the mineralized part obviously does not take place, cells inside the bone are thus provisioned.

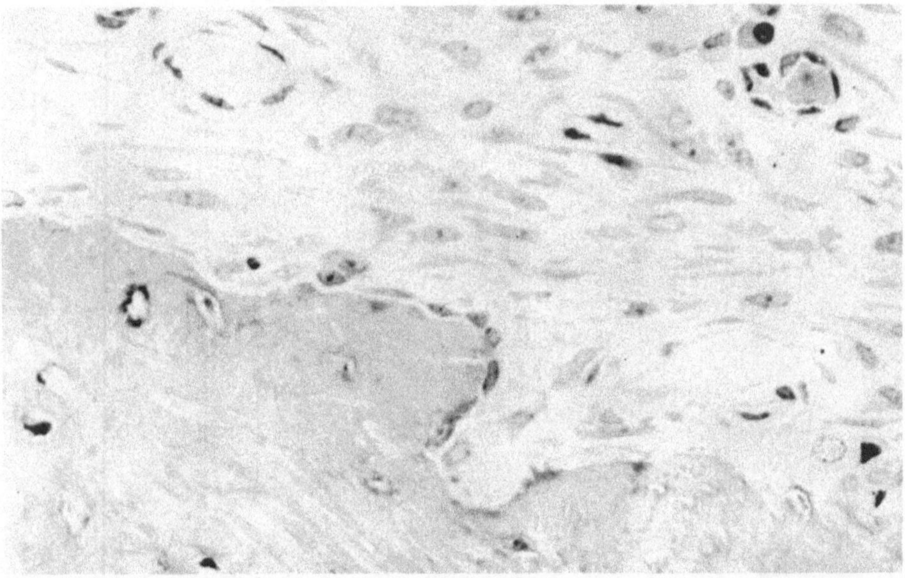

FIGURE 3. Inactive osteoblasts lining the bone. 1 μm thick section. (Toluidine blue stain; magnification × 500.)

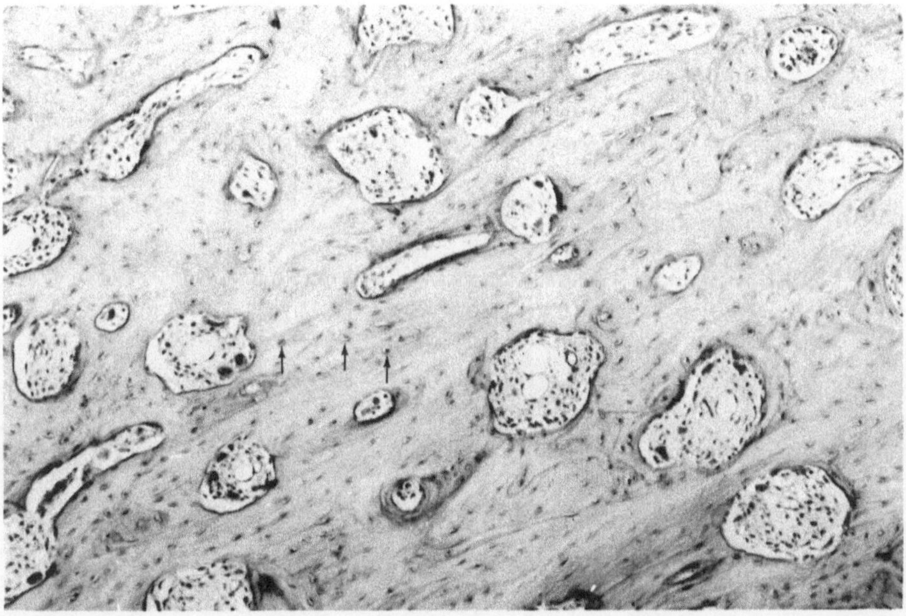

FIGURE 4. In the bone matrix numerous lacunae, each containing an osteocyte are present. Some of them are marked with an arrow. (H.E.; magnification × 80.)

D. Osteoclasts

Osteoclasts are motile cells which are able to resorb bone and are active in bone during the entire lifetime. They are found in areas where bone resorption actually takes place. In some instances just a single osteoclast may be found, while in areas where bone is resorbed rapidly, a large number of osteoclasts may be observed (Figure 6). They are easily recognizable multinucleated giant cells of greatly varying size and num-

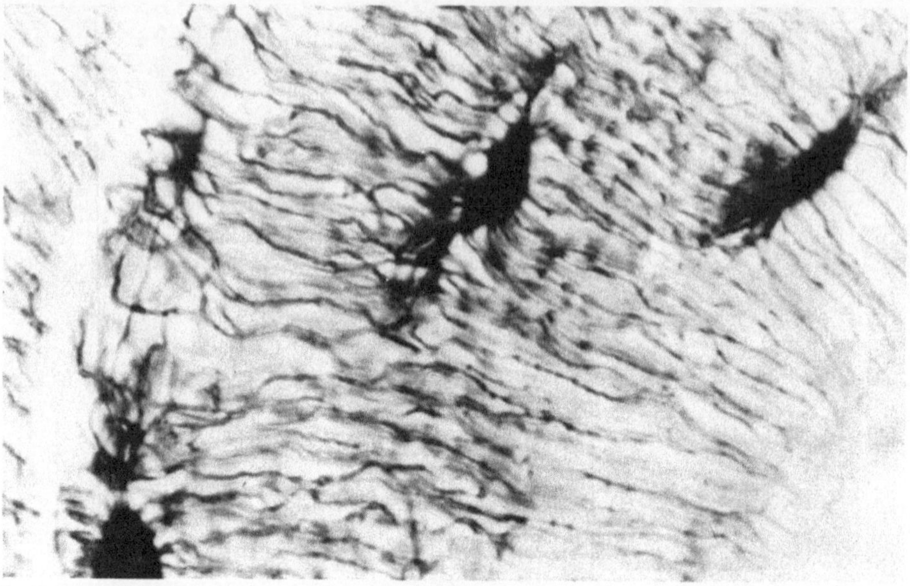

FIGURE 5. Osteocyte-lacunae with their canaculi which provide interconnections between neighboring osteocytes. (Sublimate stain; magnification × 1300.)

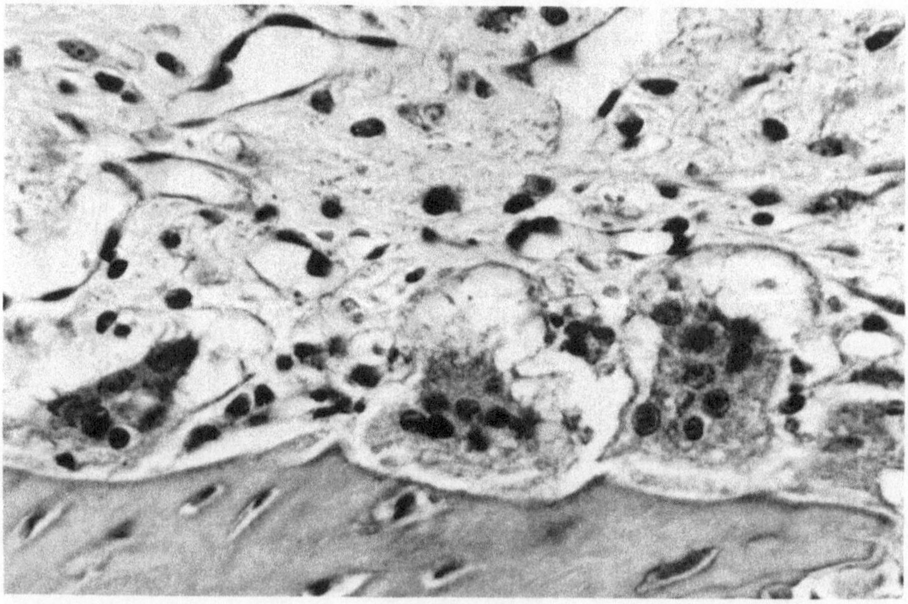

FIGURE 6. Osteoclasts lining the bone. Note the typical depressions in the bone underneath the osteoclasts, representing the resorption areas. (H.E.; magnification × 500.)

ber of nuclei (5 to 50) which are round to oval. The cytoplasm usually is eosinophilic and the vesicular nuclei contain a distinct nucleolus, but have few chromatin (Figure 7). In microscopic sections they are found at a small distance from the bone surface, apposed to a bone surface or in bays and grooves which originate from local resorption of the bone, the Howship's lacunae. They also can be found lying against chips and

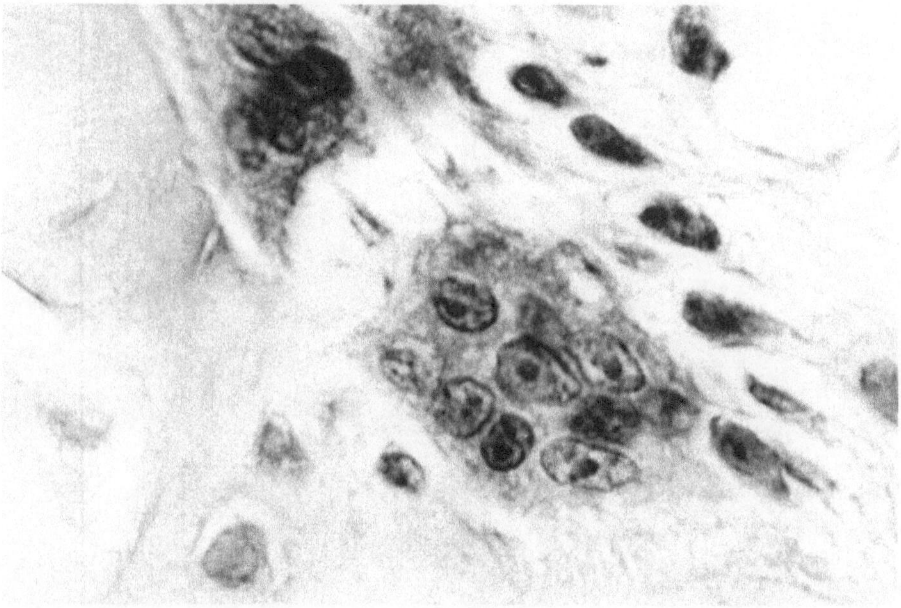

FIGURE 7. High power micrograph of an osteoclast. Several nuclei with a distinct nucleolus but with few chromatin are visible. (H.E.; magnification × 1300.)

grindings which originate from a surgical procedure, such as drilling with a bur. They are able to remove vital as well as dead (necrotic) bone. In an active osteoclast, the outer part of the cell-surface adjacent to the bone, shows a tight contact (the sealing zone) with it through the electron microscope. The central portion of this part of the cell surface demonstrates many microvilli, the ruffled border. This structural organization of the cell relates to its function.

E. Bone Matrix

The bone matrix consists of an organic and an anorganic phase and water.

The anorganic phase accounts for about 60% of the total weight of the matrix. It consists mainly of calcium and phosphate, but bicarbonate, citrate, magnesium, potassium, and sodium are also present. The calcium and phosphate are present in the form of the so-called hydroxyapatite, $Ca_{10}(PO_4)_6(OH)_2$. It forms needle-like crystals which are oriented alongside the collagenous fibers.

The organic phase accounts for about 30% of the total weight of the matrix. It consists mainly of collagen which is laid down in the form of fibers. Apart from the collagen, glycoproteins and proteoglycans are also present.

The water content of the bone matrix is about 10% of the total weight.

F. Periosteum

The periosteum covers the external surface of the bones (Figure 8), except for the surfaces of joints. It consists of two layers: an outer layer of dense connective tissue which contains fibers and blood vessels and an inner layer containing loosely arranged connective tissue fibers, spindle-shaped fibroblasts, osteoprogenitor cells, and osteoblasts (Figure 9). When the osteoblasts are in an inactive state (Figure 10), as is frequently the case in the adult, the osteoblasts are indistinguishable from the fibroblasts and the osteoprogenitor cells. Such a periosteum has no osteogenic function, but has osteogenic potential. For example, during remodeling after bone fracture, the bone

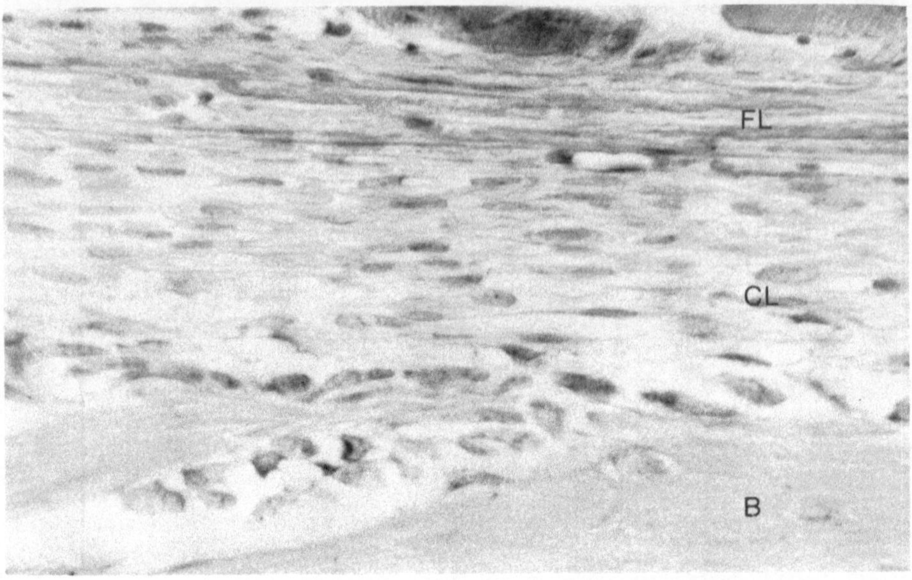

FIGURE 8. Periosteum covering the external surface of a long bone (B). An outer fibrous layer (FL) and an inner cellular layer (CL) can be distinguished. (H.E.; magnification × 500.)

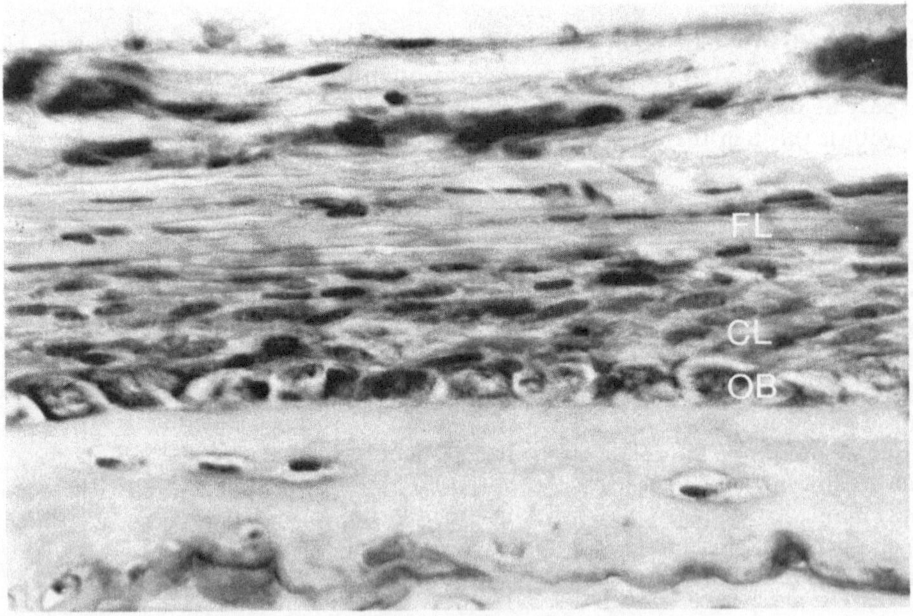

FIGURE 9. Active periosteum with a continuous layer of active osteoblasts (OB). The fibrous layer (FL) and the cellular layer (CL) are clearly distinguishable. (H.E.; magnification × 500.)

matrix-forming potential becomes effective and active osteoblasts can be observed adjacent to the bone. These osteoblasts are derived from the osteoprogenitor cells in the periosteum. In the young organism osteoblasts are present in the inner layer of the periosteum and osteogenesis takes place by apposition by cells of interstitial substances. Usually, the periosteum is tightly anchored to the bone by collagen fibers from the outer and inner layer which insert into the bone. That part of the fiber being incorporated in the bone is called Sharpey's fiber.

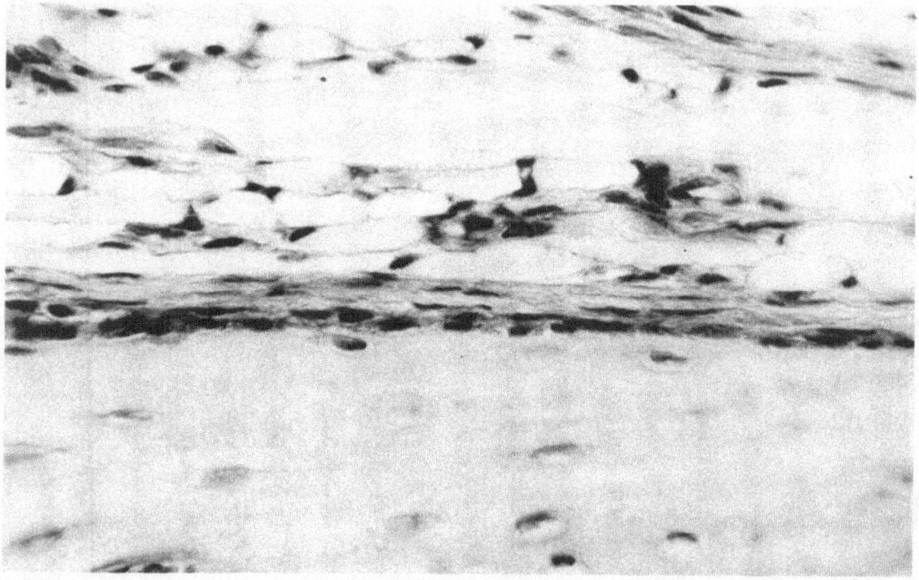

FIGURE 10. Resting periosteum containing inactive osteoblasts or osteoprogenitor cells in the cellular layer. (H.E.; magnification × 500.)

G. Endosteum

The endosteum usually is a thin layer of connective tissue containing a few cells lining the inner surface of bones: the compacta of long bones or the spongy bone. In the endosteum no fibrous layer is present as can be observed in the periosteum (Figure 11). Osteoblasts are not recognizable, except in areas of bone growth or after the bone has been traumatized by fracture or by surgical procedures. In such cases, these active cells originate from osteoprogenitor cells in the endosteum. In these cases the osteogenic potential of the endosteum is evident.

III. BONE DEPOSITION AND RESORPTION

A. Introduction

Both formation and resorption of bone tissue are simple processes morphologically. They play a role as well in direct as in indirect bone formation and are identical in both types of bone development. The formation always takes place by apposition. New bone is laid down whether in soft connective tissue, on bone itself, or on cartilage. In all cases, the osteoblast is the leading cell.

Resorption is always a cellular activity as well. Most commonly bone resorption is performed by osteoclasts which, when active, lay against the surface of a bone. Moreover, osteocytes may be engaged in bone resorption, the so-called osteocytic osteolysis. Both processes are relevant not only during the growth of bones (which is accompanied by internal reconstruction of the bone, remodeling), but also after the bone has acquired its final size and shape, in which case remodeling continues. Moreover, both processes are relevant in fracture healing and after surgical procedures in which bone is engaged. This is true even when the cells have become inactive prior to the noxe. Subsequent to the noxe, the cells can be reactivated, demonstrating their osteogenic potency (osteoblasts), or their resorption potential or originating anew from their respective cells of origin.

B. Bone Deposition

Formation of (new) bone, is primarily an activity of osteoblasts. They differentiate

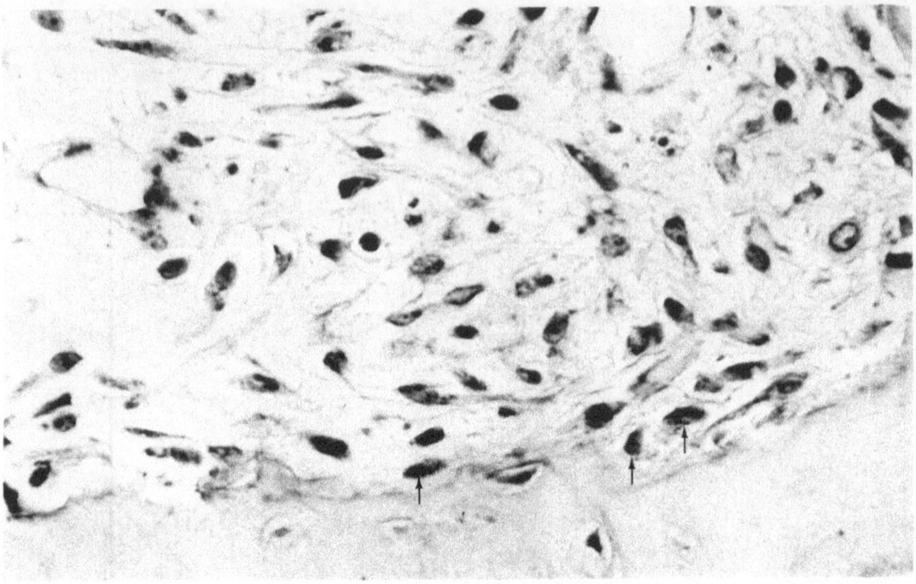

FIGURE 11. Active endosteum in which only osteoblasts can be distinguished (arrows). No fibrous layer is present. (H.E.; magnification × 500.)

via the osteoprogenitor cells after mitosis from embryonic (undifferentiated) mesenchymal reserve cells which are usually present in the soft connective tissue. Functioning osteoblasts can also arise from inactive osteoblasts in a resting periosteum or endosteum. In most cases, a continuous layer of osteoblasts is arranged adjacent to the bone. These cells produce the organic intercellular substance (matrix) of bone. It consists of collagen fibrils and fibers which are luted with glycoproteins and proteoglycans. This organic intercellular substance (the osteoid) can be observed in the form of a seam in some of the depository areas. The seam stains bright red after H.E. staining (Figure 12). During osteoid formation some osteoblasts become embedded in the osteoid. The periosteum or endosteum is replenished with osteoblasts which are differentiated from mesenchymal reserve cells via the osteoprogenitor cell. The osteoid is not yet mineralized.

The mineralization of the osteoid lags behind osteoid deposition.

The relative speed of mineralization and osteoid formation determines the width of the seam. Thus, no seam or one that is narrow can be discerned. It may even be a wide seam. The calcium and phosphate originate from the blood. Some participation of the osteoblast in the mineralization is postulated. Mineralization is attended with the intracellular activity of the enzyme alkaline phosphatase. However, mineralization is far from fully understood. The embedded cells which are now called osteocytes, gradually lose their matrix forming activity. The lacuna in which the cell body of the osteocyte fits snugly in young bone is small in mature bone.

When bone formation stops, the osteoblasts become inactive. They lose their distinct stainability for RNA and become flattened and can dedifferentiate to osteoprogenitor cells. Such cells are found apposed to "resting" bone surfaces where neither deposition nor resorption takes place.

C. Bone Resorption

Osteoclastic bone resorption is performed by multinucleated giant cells which arise from fusion of several osteoblasts, from undifferentiated mesenchymal cells undergoing differentiation to osteoclasts, or from differentiation from monocytes.

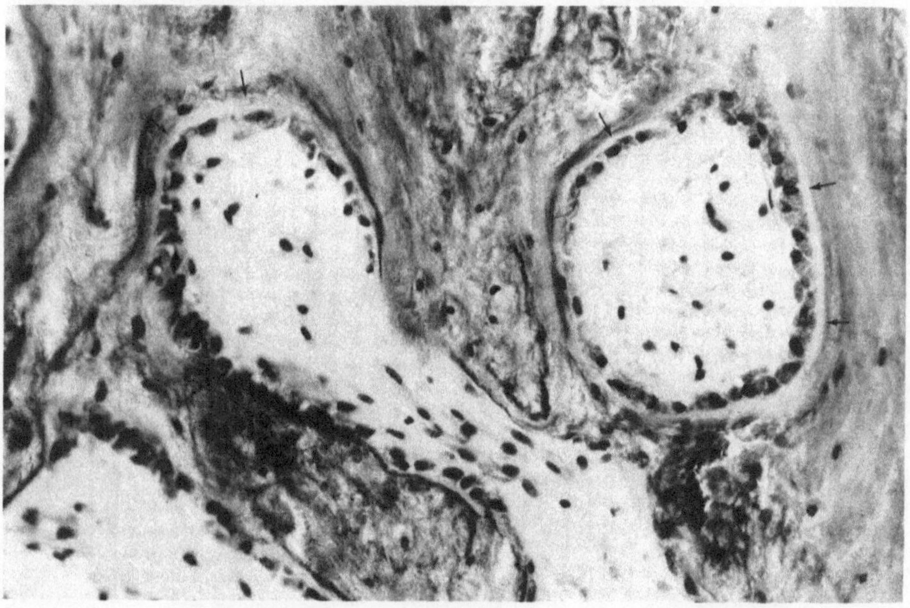

FIGURE 12. Osteoid seam (arrows) between endosteal layer of osteoblasts and the mineralized bone. (Hematoxylin-toluidine blue-acid fuchsin stain; magnification × 200.)

Stimuli which provoke osteoclastic activity vary from mechanical forces (e.g., the mere mechanical reaming of the root canal during endodontic treatment where bone resorption is provoked at the lamina dura) hormones, vitamins, and the presence of overaged or necrotic bone, to the presence of an inflammatory reaction. The final chemical substance or physical influence which leads to the appearance of osteoclasts is not known. However, an osteoclast-activating factor is isolated which, for example, plays a role in inflammatory reactions.

The osteoclast is able to secrete acid into the space between the cell membrane of the ruffled border and the mineralized bone. Thus, a superficial layer of mineral opposite the ruffled border is dissolved. In appropriate sections where the osteoclast is observed in a Howship's lacuna, an osteoid-like seam can be visualized at the bottom of the lacuna (Figure 13). Then the osteoclast secretes proteolytic enzymes, one of which is collagenase. These enzymes are capable of degrading the ground substance and collagen fibers, thus providing the complete removal of osseous materials. Acid and enzymes are believed to be confined to the extracellular space which is limited by bone and the osteoclast with its sealing zone.

Osteocytic osteolysis can be performed by the osteocytes located in their lacunae in the bone. It is a type of resorption from within the bone. The process of osteocytic osteolysis plays a role in the homeostasis of calcium or can lead to the death of the cell involved. The mineral content of the bone matrix surrounding the lacuna decreases while the amount of organic material decreases at the same time. The walls of the lacunae demonstrate irregularities and the cytoplasm becomes hypertrophic. Thus, the cell body is enlarged. Subsequently, the cell further degenerates and dies. Quantitatively, the disappearance of bony mass by osteocytic osteolysis is small when compared to the amounts which are removed by osteoclasts. This is both in normal and pathological conditions.

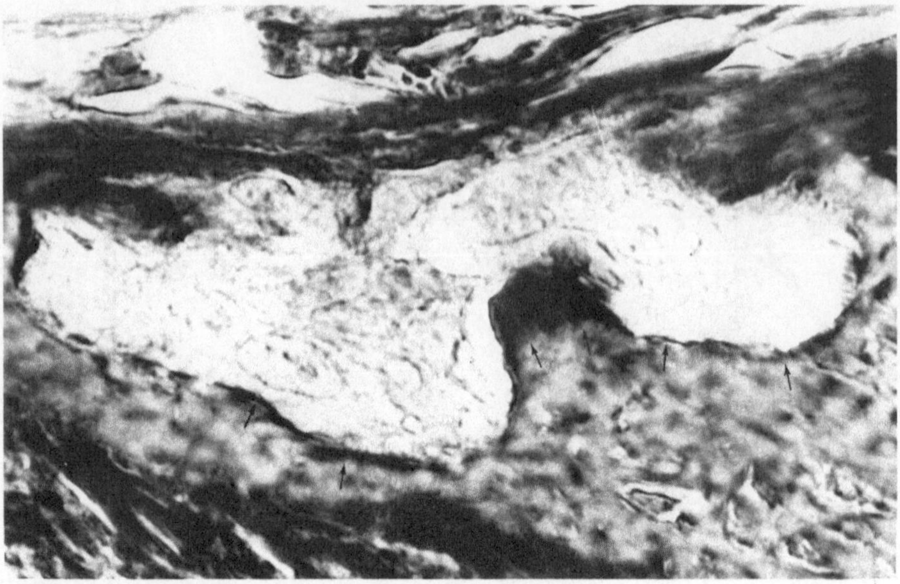

FIGURE 13. Two osteoclasts in Howship's lacunae. Between the cells and the mineralized bone marrow seams and an area of blue staining demineralized matrix is observed (arrows). Lamina dura of a monkey. (Masson trichrome stain; magnification × 1300.)

IV. TYPES OF BONE

A. Introduction

Macroscopically, two types of bone can be distinguished:

1. Compact osseous tissue which is found in the shafts and heads of long bones, in the cortex of flat bones (e.g., in the skull) and irregular bones (e.g., the carpus)
2. Spongy bone (cancellous bone) which is present in the heads of long bones, the flat bones (cranium), and in all irregular bones (e.g., vertebrae, carpalia, and facial bones)

Microscopically, two types of bone can also be observed. This distinction is based on the orientation of the collagen fibers within the bone:

1. Woven bone, in which the collagen fibers run in three directions
2. Lamellar bone, in which the collagen fibers show a regular orientation in the form of lamellae.

The above-mentioned appearances would suggest that a total of four types of bone are in existence. However, compact woven bone does not exist.

In the rapidly growing juvenile, some bones primarily consist of woven bone, showing a spongy character. During further growth, lamellar bone is deposited against this woven bone. In the adult, almost no woven bone is present because of the continuous remodeling. The following discussion is directed to the adult situation, thus dealing with lamellar bone, although some remarks will be made on the woven bone which is formed during repair processes.

B. Spongy Bone

Cancellous or spongy osseous tissue can be present in the medulla of compact bones.

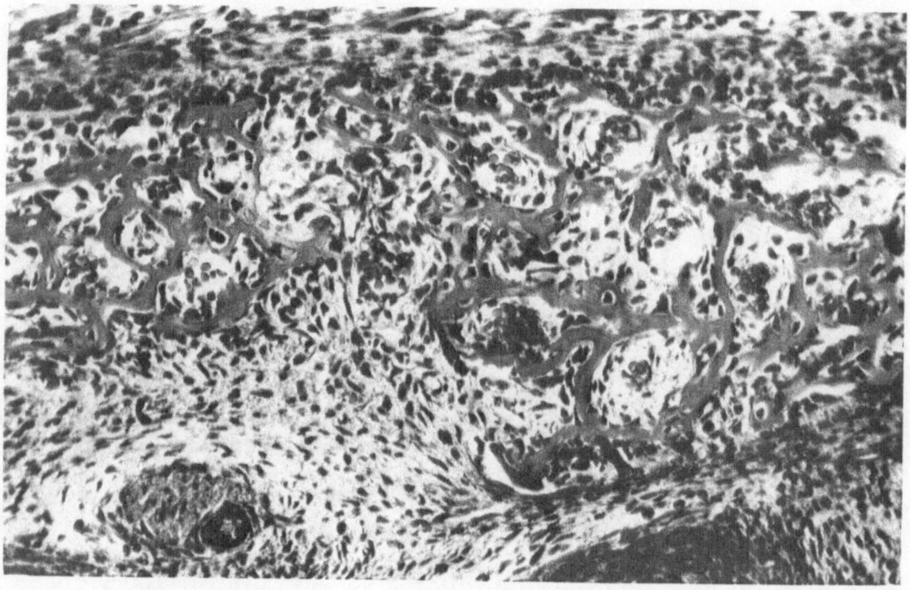

FIGURE 14. Spongy bone covered with numerous osteoblasts indicating a rapid bone deposition. (Masson trichrome stain; magnification × 200.)

In this case, it is surrounded by a layer of compact bone. The spongy bone can be observed as a network of bars and plates (Figure 14), the trabeculae, which overall demonstrates a well-defined orientation in the normal situation. The trabeculae at each location in the bones function in such a way that they transfer mechanical forces from one part of the surrounding compact osseous tissue to another part. The trabeculae for the greater part, run in directions corresponding with the lines of pressure or tension.

The trabeculae are composed of thin layers of interstitial substance (lamellae) in which collagen fibers run. The osteocytes are located in the contact surfaces of the lamellae and they communicate by means of their canaliculi. In the spaces between the trabeculae, bone marrow is found. It is either red bone marrow, which mainly consists of blood forming cells (hematopoietic tissue), or yellow bone marrow, mainly consisting of fat cells.

Primarily, spongy bone is of the woven type. During development it is gradually replaced by spongy bone of the lamellar type. New lamellar bone is deposited at the original trabeculae as well. The functional significance of this resorption-deposition process is two-fold: adaptation to changing mechanical loads and to obtain maximal strength with minimal weight.

When a cavity is drilled into spongy bone and after the introduction of an implant, an initial resorption of trabecular bone is observed around the implant. The amount of resorption is dependent upon the kind of the implant material and the surgical procedure. The resorption can vary from practically zero, as for example, with some ceramics and hydroxyapatite combined with careful surgery, to enormous amounts when, for instance, too much heat is produced during drilling and a noxious agent leaks out from the material. However, the resorption is followed by deposition of new bone when the surrounding tissues are no longer threatened, thus providing a close fit of the implant. The new bone which is continuous with the spongy bone, is more compact than the surrounding spongy bone, thereby resembling the lamina dura of a tooth socket. Whereas the lamina dura is separated from the tooth by the periodontal

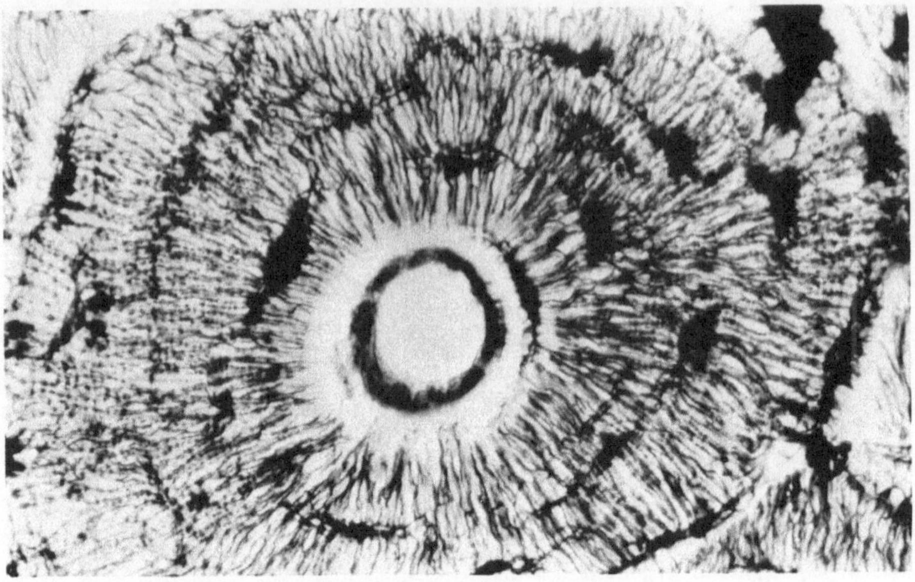

FIGURE 15. Osteon showing its lacunae in circular arrangement. These circles depict the boundaries between the lamellae of the osteon. The haversian canal is centrally located. (Sublimate stain; magnification × 500.)

ligament, the newly deposited bone locally is immediately adjacent to such implants or at a slight distance from it. This phenomenon, together with other factors, can provide a firm anchoring of the implant.

C. Compact Bone

Compact bone which, as an example, is found in the shaft of the long bones, is composed of lamellae which are luted together, thus resembling the structure of plywood. The osteocytes are located between the lamellae. They are produced by the periosteum and the endosteum, and consequently are found at the external and internal surfaces, respectively, of the shaft of long bones.

The architecture of compact bone is complicated by the so-called haversian systems. During the early development of compact bone, lamellae are deposited by the periosteum. In these lamellae, tunnels are saved in which blood vessels are present. These provide for the vascularization of the osseous tissue itself and of the marrow.

When the layer of the periosteal lamellae is of a certain thickness, osteoclasts resorb parts of these lamellae, thus creating tunnels within. These tunnels will eventually contain blood vessels and are called haversian canals. They are much wider than would be necessary to lodge the blood vessels. Cells of the connective tissue surrounding the blood vessels differentiate into osteoblasts and these deposit bone lamellae on the wall of the canals (the haversian lamellae which obviously are concentric). This process stops when the canal is just wide enough to lodge the two (there may be more) blood vessels and a few connective tissue. This system of lamellae, its cells, and the canal is called the osteon.

The cells are located in lacunae in the bone and intercommunicate with their processes which run in canalicules. The innermost canalicules open into the lumen of the haversian canal. In this way, tissue fluids, originating from the blood, can reach even the outermost osteocytes. Thus, the life of the osteocytes is maintained. The osteocytes are arranged in concentric circles in the bone tissue (Figure 15). In the shaft of long bones osteons are found in the inner ⅔ of the thickness of the wall. The outer ⅓

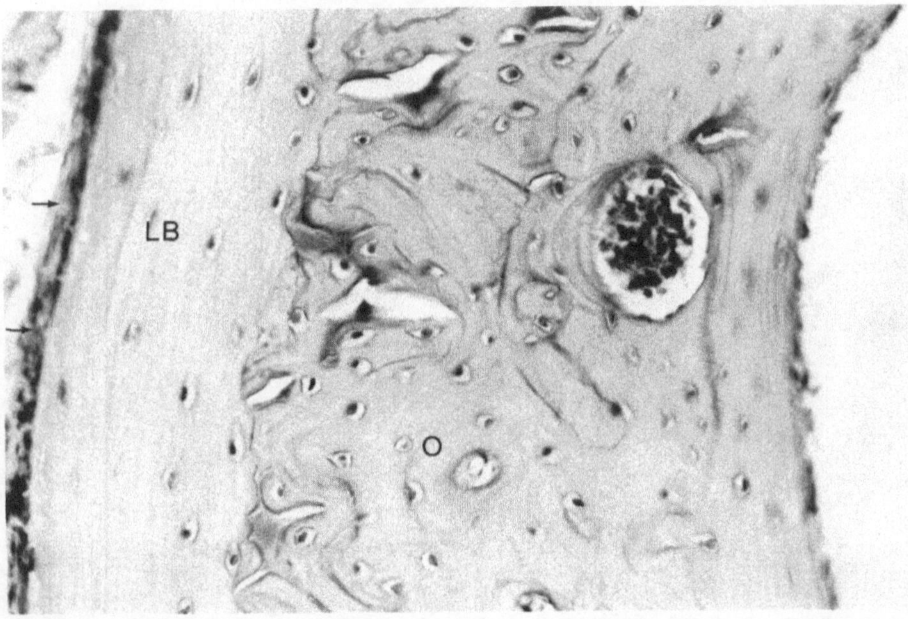

FIGURE 16. Transverse section through the diaphysis of a long bone. At the external surface a periosteum is present (arrows). The lamellar bone in the outer part of the shaft (LB) is deposited by the periosteum. More internally a zone is found in which osteons are present (O). At the internal surface again some lamellar bone is found (H.E.; magnification × 325.)

consists of basic or circumferential lamellae which originate from the periosteum. At the internal surface a small number of inner basic lamellae are found which are deposited by the endosteum (Figure 16).

In the shaft of long bones, the haversian canals run length-wise. Transverse canals are found as well and are called canals of Volkmann. They interconnect the haversian canals mutually and with the external tissue and bone marrow.

The collagen fibers present in each lamella of an osteon run parallel to each other and spirally to the canal. In adjacent lamellae the fibers cross alternately at various angles. This mode of orientation adds to the strength of the osteon and is also responsible for the alternately bright and dark lamellae and the Maltese cross which can be observed in cross-sections of an osteon in the polarizing microscope.

Regarding the structure of compact bone, another complicating factor should be mentioned. After construction, osteons can be resorbed by osteoclasts and can be rebuilt afterwards. However, resorption is neither limited to the original boundaries of the osteon nor complete for that osteon. Consequently, after the formation of a new osteon, remaining lamellae of old osteon(s) can be observed, filling angular spaces between haversian systems. These lamellae are called interstitial or ground lamellae and are clearly seen in transverse sections of long bones. Evidently, the location of the blood vessel is changed at the same time.

The continuous resorption and construction of haversian systems can be considered a mode by which adaptation of the bone to changing mechanical loads is possible, as osteons increase the strength of bone.

D. Woven Bone

The collagen fibers in woven bone run in all directions. This accounts for the fact that in the polarizing microscope a distinct orientation of the fibers is not observed. Woven bone is also called embryonic bone or immature bone. It possesses a great

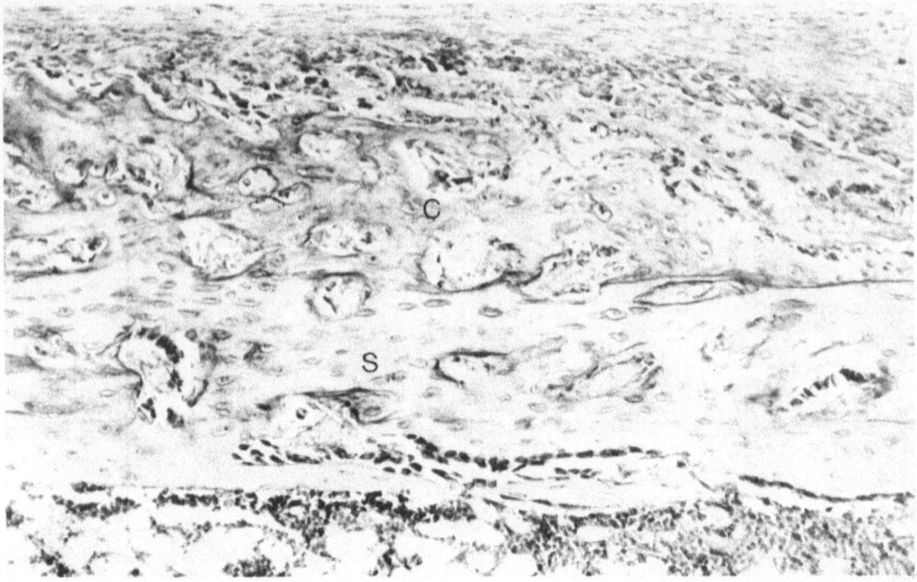

FIGURE 17. A callus (C) is deposited at the external surface of the shaft (S) of a long bone. The callus is spongy bone of the woven type, while the shaft mainly consists of lamellar bone. (H.E.; magnification × 125.)

number of relatively large lacunae which are filled with the cell bodies of osteoctyes. The lacunae are haphazardly distributed over the matrix, contrary to the normal lamellar spongy bone in the adult.

Woven bone is also formed during adult life after fractures or tooth extractions. Afterwards, it is replaced by lamellar bone. During the healing of the fractures, the formation of woven bone that usually mineralizes is accompanied by cartilage formation in the majority of the cases. This combination is called the callus (Figure 17). The great number of cells and thus the smaller amount (at least initially) of lesser mineralized substance per unit volume of woven bone, accounts for the poor roentgenographic detectability. Even in the case that it can be histologically demonstrated that the union of the fragments of a fracture is established or the extraction socket is filled with woven bone, it can be invisible in radiographs.

V. HISTOGENESIS AND REMODELING OF BONE

A. Introduction

Formation of bone is not possible when a good vascularization and a reactive connective tissue are not present. A good blood supply is necessary with regard to synthetic activity of the osteoblasts and mineralization. A reactive connective tissue is not present in a young individual only. In the case a sufficient pool of undifferentiated mesenchymal cells and/or fibroblasts are present, bone formation can take place in the adult as well (fracture repair, bone growth around and into implants).

Bone develops not only by (the cells of) connective tissue, but also within connective tissue. Roughly, the steps in the process leading to bone are the development of osteoblasts from undifferentiated mesenchymal cells or fibroblasts, the formation of a matrix by these cells which is mineralizable, and the mineralization itself. These steps are equal for both types of bone formation: the direct and the indirect ossification. Direct or intramembranous or desmal bone formation takes place without the previous existence of a mold. Indirect or intracartilaginous or endochondral ossification starts on the basis of a cartilaginous mold which approximately determines the final shape of the bone. The model is removed during the process of ossification.

In the following section, both types of ossification will be described in the order given. The histogenesis of the cartilaginous model, in the case of the indirect ossification, falls beyond the scope of this chapter.

In addition to these processes of ossification which are both associated with the skeletal system, some remarks will be made regarding ectopic bone formation. This usually abnormal bone formation takes place at locations outside the skeletal system.

B. Direct Bone Formation

Desmal ossification appears directly, i.e., without a preceding phase of cartilage formation. Several bones of the skull, e.g., the calvarium, the maxilla, are formed in this way. Direct bone formation takes place in mesenchyme or reactive connective tissue.

In the well-vascularized mesenchyme which covers the brains of a human embryo of 8 to 10 weeks, one or two ossification centers develop per final skull bone. An ossification center arises from a group of mesenchymal cells which lose their normal spindle shape and become polyhedral. Moreover, the vascularization locally is increased at the same time. At this stage of differentiation the cells are called osteoblasts and are connected with one another by long cellular processes. The osteoblasts deposit a dense intercellular substance between them. Some cells become embedded in this matrix while others stay at the periphery of the area. The dense matrix contains thick collagenous fibers running in all directions. This is contrary to the interstitial substance which surrounds the center of ossification. After mineralization, the ossification center contains bone (Figure 18).

During further development of the skull bone, three processes can be distinguished:

1. The ossification spreads out in the connective tissue through neighboring connective tissue becoming involved in the process of ossification described above. Thus a plate of bone of a spongy type is affected in which the ossification centers of one skull bone are incorporated. In this way the areas of connective tissue between the growing bones become smaller (Figure 19). Ultimately these are the sutures which are still fairly large (especially the fontanelles) in the newborn.
2. At the same time the bone increases in thickness. It occurs in the way described above at first, but later the connective tissue above and below the ossification center takes on a membranous character. A periosteum and an endosteum are formed. The periosteum which envelopes the skull bone, consists of an inner cell-rich (osteoprogenitor cells and osteoblasts) layer, also containing collagenous fibers, and an outer fibrous layer. In the latter layer, the fibers run parallel to the surface of the bone. During progressing ossification, cells and fibers become enclosed in the matrix produced by the osteoblasts. This so-called periosteal (endosteal) ossification gives rise to lamellar bone. The used cells are supplied from outside connective tissue.
3. Growth of the embryo necessitates remodeling. With enlargement of the skull, the curvature decreases. Hard tissue cannot adapt to this without cellular activity. Generally, externally more periosteal lamellar bone is apposed, whereas internally, osteoclasts resorb bony tissue. The resorption is intermittent, lamellar bone being deposited in the meantime to keep the internal surface of the skull bone smooth. In the course of time all original, nonlamellar, woven bone is resorbed. As remodeling continues, only lamellar bone is resorbed and deposited. Ultimately skull bones contain a diploë: a spongy core which is enclosed by a tabula

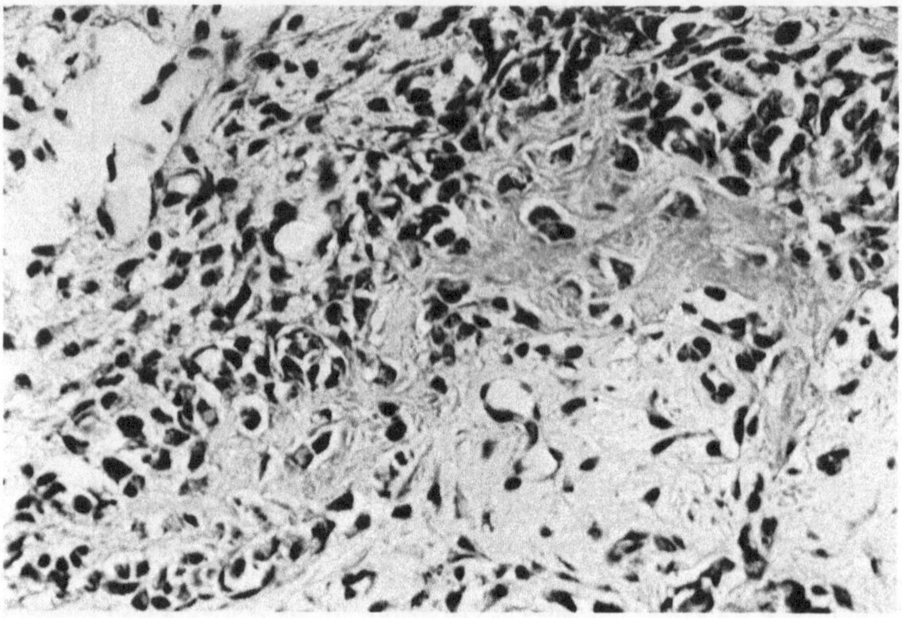

FIGURE 18. Ossification center. Some bone is deposited and some osteocytes are already found in the matrix. The bone is lined with numerous osteoblasts. (H.E. magnification × 325.)

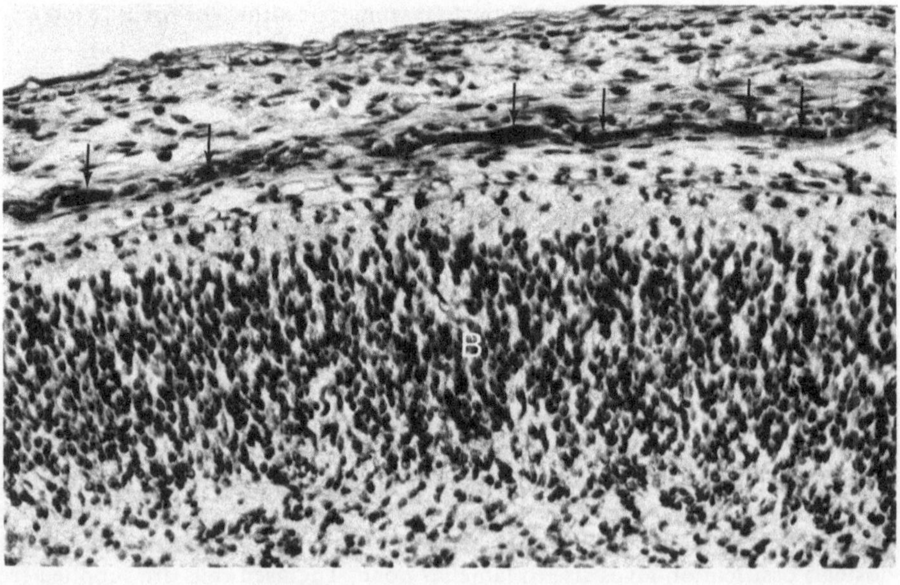

FIGURE 19. Ossification centers (arrows) in an embryonic cranial bone. The black areas represent the already mineralized bone matrix. The developing bone is covered by fibrous envelope. Brain tissue (B) is found underneath the bone. (Von Kossa stain; magnification × 200.)

externa and a tabula interna which both consist of compact lamellar bone. The diploë is the result of partial resorption of lamellar bone and therefore is unlike the initial spongy bone. The cavities of the spongy bone contain myeloid tissue (red bone marrow) which persists through adult life.

C. Indirect Bone Formation

Indirect bone formation occurs during embryonic life at places where the body initially constructs a mold of hyaline cartilage.

The bones of the thorax, the limbs, and some bones of the skull develop on the basis of a hyaline cartilage model. In this case also, the process of ossification starts with well-vascularized reactive connective tissue. Cartilage itself never changes to bone and chondrocytes never pass into osteocytes. The solid cartilage mold of a long bone shows a shaft. This is the diaphysis in the final situation and is located between two condyles or the epiphysis. This model is surrounded by a perichondrium which has a structure similar to that of a periosteum.

Around the center of the cartilaginous diaphysis, the perichondrium will behave as a periosteum, i.e., it forms a lamellar spongy bone and consequently can be called periosteum from this moment. In time, this periosteal ossification stretches in proximal and distal directions and stops when the condyles are reached. This bony sleeve is called the periosteal bone band or collar.

Shortly after the start of the formation of the bone collar, the endochondral ossification begins. The chondrocytes in the center of the cartilaginous diaphysis swell, at the expense of the interstitial tissue, and degenerate. The interstitial tissue mineralizes. Then, blood vessels and connective tissue penetrate the transformed cartilage at a number of places. Chondrocytes and most of the interstitial tissue are removed by chondroclasts. Mesenchymal cells that have penetrated differentiate into osteoblasts that deposit bone on the cartilaginous remains. When this process has proceeded in the direction of the epiphysis, the hard tissue in the diaphysis is removed so that the periosteal bone collar surrounds a cavity (bone marrow cavity) which is mainly filled with soft tissue. This tissue develops into fatty tissue and blood-forming tissue. The process of cartilage removal is limited to the content of the periosteal bone band.

In the epiphysis, ossification is mainly of the endochondral type as well. In the majority of the long bones, the ossification of the epiphysis has just begun in the newborn. In starts centrally after the penetration of blood vessels and connective tissue into the cartilage. The ossification spreads radially in a way which is similar to that described for the growing shaft of the long bone. When the growth of the epiphysis has ended, the progressing ossification reaches the surface, except at the joint surface where cartilage is left. This is in direct contact with the synovial fluid. At the remaining surface, the perichondrium passes into a periosteum which deposits a layer of bone upon the spongy bone of endochondral origin. The periosteal bone later changes into compact bone. At the side of the diaphysis a bone plate is formed (epiphyseal bone plate).

Shortly after the start of ossification in the epiphysis, chondrocytes of the mold present in a zone between epiphysis and diaphysis start dividing and start the formation of new cartilaginous intercellular substance. Mitoses take place in such a way that after a few divisions, a cartilaginous zone between diaphysis and epiphysis has arisen in which the rows of chondrocytes run parallel to the length-axis of the bone. This zone of epiphyseal cartilage, with its mitoses and formation of intercellular substance, accounts for the linear growth of the long bone (Figure 20). The growth of the cartilaginous zone proceeds with equal speed as the ossification until growth ceases.

In a diaphyseal direction the following zones of ossification can be observed. The first zone consists of swollen chondrocytes. Vacuoles are formed in the cytoplasm. The nuclei lose their chromatin. This zone of degenerating cells is called the zone of the vesicular of hypertrophic cartilage.

The second zone is that of mineralization of the bands of interstitial substance between the columns of degenerated chondrocytes, the so-called zone of provisional or preliminary mineralization. The mineral in this rather narrow zone reinforces the junction between the hyaline cartilage of the epiphyseal plate and the spongy bone of the periosteal bone collar.

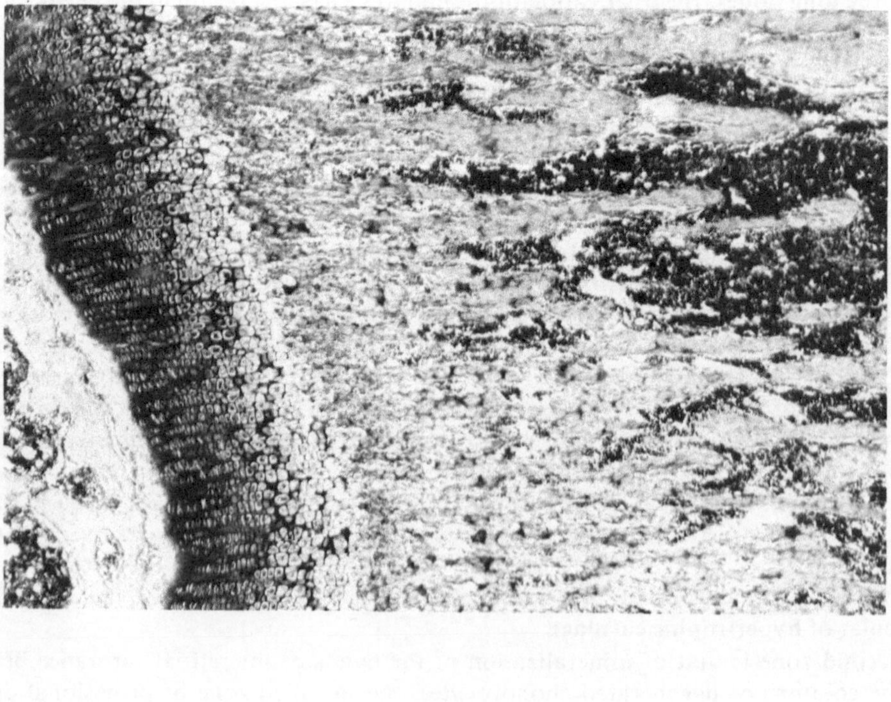

FIGURE 20A. Low power micrograph showing an epiphyseal plate. (H.E.; magnification × 50.)

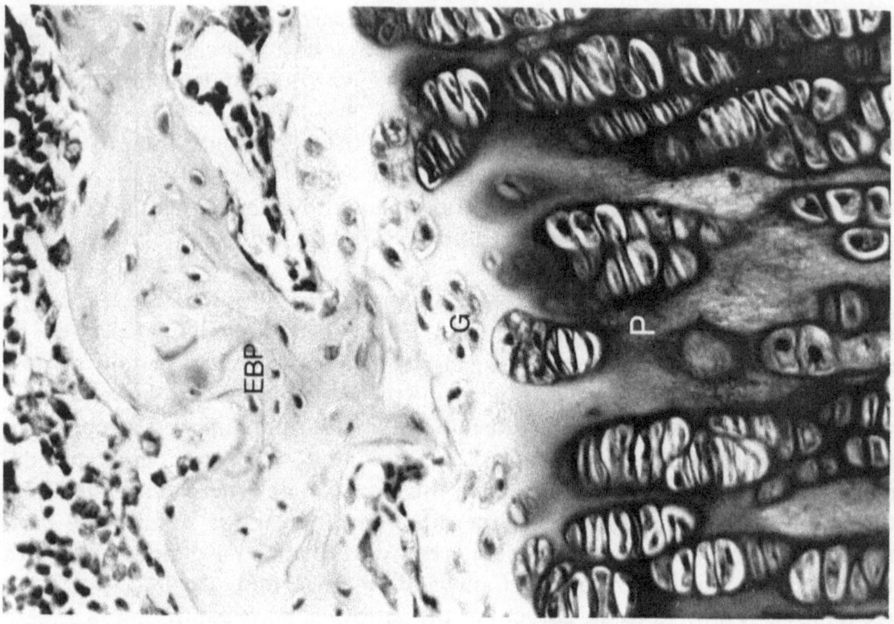

FIGURE 20B. Detail of an epiphyseal plate showing the epiphyseal bone plate (EBP), the germinative zone (G), and part of the proliferative zone (P). (H.E.; magnification × 325.)

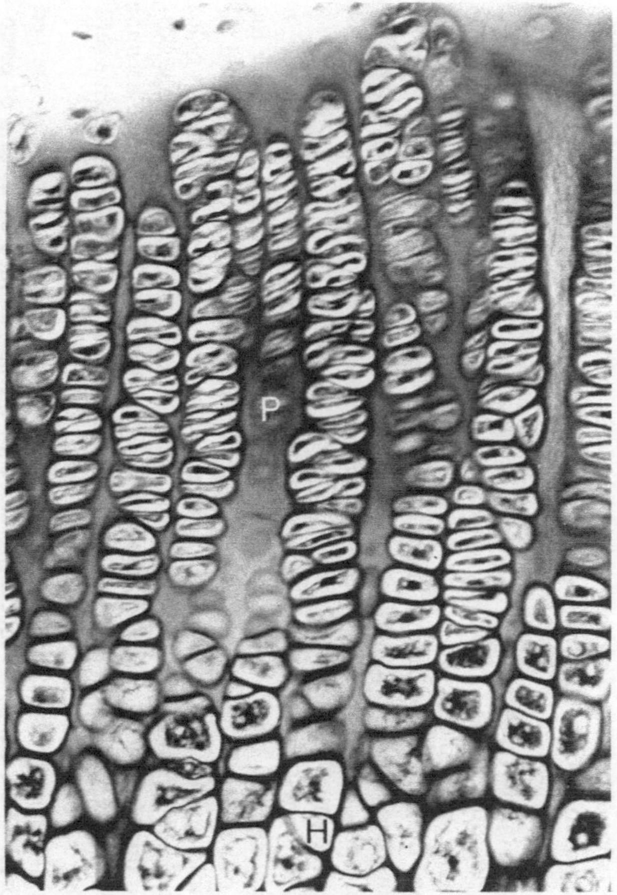

FIGURE 20C. Detail of an epiphyseal plate showing the proliferative zone (P) and the hypertrophic zone (H). (H.E.; magnification × 325.)

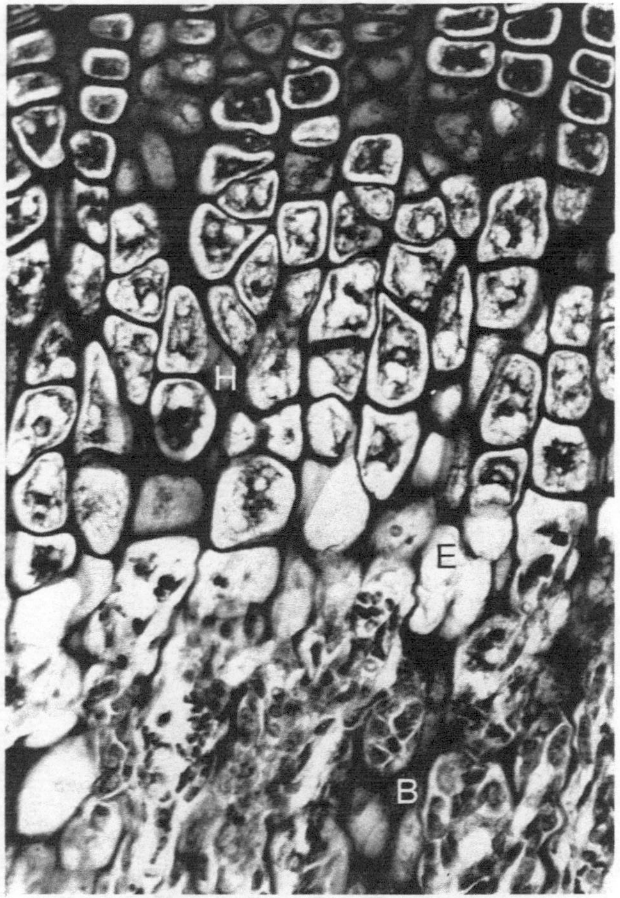

FIGURE 20D. Detail of an epiphyseal plate showing the hypertrophic zone (H), the erosion zone (E), and the zone of bone formation (B). (H.E.; magnification × 325.)

The third zone shows the penetration of blood vessels and connective tissue from the marrow in the shaft into the rows of chondrocytes. For this penetration the chondrocytes and the walls of interstitial substance between the chondrocytes in the columns are degraded. However, the mineralized interstitial substance between the columns of cells usually is not decomposed. Thus, canals originate that may communicate locally and that contain blood vessels and connective tissue. The osteogenic potential of the connective tissue is demonstrated by the fact that osteoblasts can be observed on the mineralized canal walls. These cells start producing matrix which normally mineralizes shortly after its production. In this way, locally, mineralized cartilage covered by mineralized bone matrix that is covered by a seam of osteoid (usually in a thickness of about 1 μm) to which a layer of osteoblasts is located, can be observed. This zone is called the metaphysis.

The metaphysis is steadily remodeled. One of the ways in which it is remodeled is by osteoclasts that resorb the tips of the trabecular columns. During the period, the rate of growth is constant; the length in the complex of the nonmineralized epiphyseal cartilage disk, the zone of preliminary mineralization, and the zone of mineralized trabeculae (as they appear in microscopical sections) are also constant.

At the end of the growth period (at the age of 20 to 22 years in the human) cell division and formation of interstitial substance in the epiphyseal cartilage plates comes to an end. The epiphyseal cartilage plates are completely replaced by bone, the so-called closure of the epiphyseal plates.

Growth in thickness of the shaft of long bones after the initial periosteal bone formation takes place similar to that in the direct ossification. The periosteum deposits lamellar bone. At the sites where blood vessels penetrate the shaft, openings in the lamellae are maintained.

The early growth in thickness of the periosteal bone collar takes place in such a way that the initially lamellar spongy bone is replaced by compact bone by remodeling. This occurs in such a manner that externally new lamellae are formed, while internally bone is removed by osteoclasts at the same time.

D. Ectopic Bone Formation

In the preceding sections, processes have been described in which bone formation is directly associated to the skeletal system. The way in which the induction of bone formation, i.e., the factors provoking mesenchymal cells to differentiate into matrix-producing osteoblasts and its subsequent mineralization, takes place during embryonic life is only poorly known. After being induced, bone is always deposited on bone or on (mineralized) cartilage.

In ectopic bone formation, ossification takes place in a location not associated with the skeletal system. It may be a spontaneous ossification or an ossification after the addition of some irritants. Spontaneous ectopic bone formation has been described in, for example, the kidney, the walls of arteria, muscles, tendons, and the dental pulp. A well-known example of a stimulated ossification is the alcohol-induced bone formation in muscles. In all cases the locally present connective tissue is held to account for the bone formation. Undifferentiated, normally slumbering connective tissue cells have potential to differentiate to osteogenic cells.

In histological sections of porous implants, islands of bone tissue can be observed in the pores. These islands, especially when they are surrounded by a seam of osteoblasts and loose connective tissue, might give the impression of ectopic bone formation. However, serial sections of cases where porous PMMA was either implanted into a cavity in bone or was applied externally on bone (as augmentation) revealed that in all cases investigated, the "islands" appeared to be merely sections of bone trabeculae which were continuous with the bone adjacent to the implant or augmentation. The

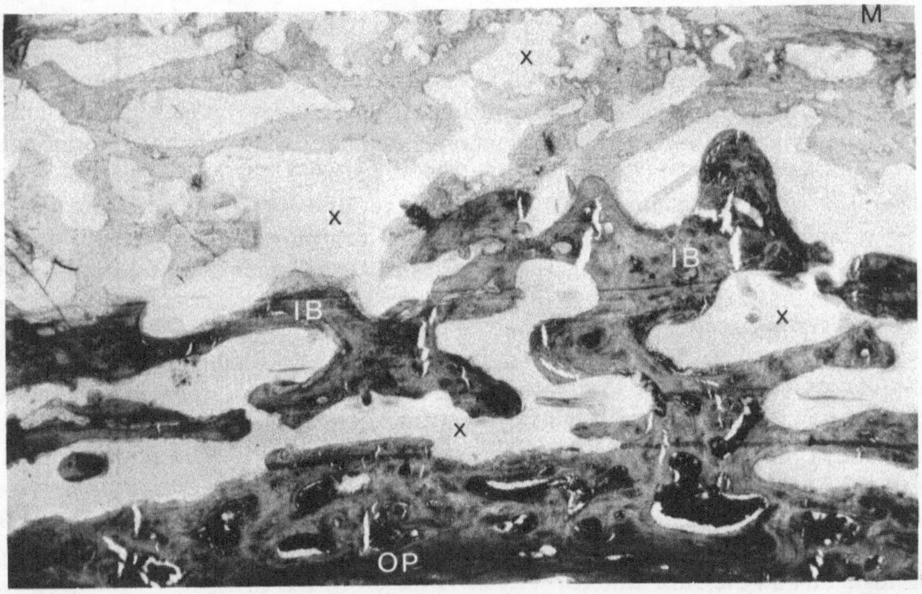

FIGURE 21. Low power photomicrograph of an augmentation on a monkey's os parietale (OP) of which the tabula externa was removed. The porous PMMA biomaterial (x) is covered by a layer of muscle cells (M). The pores are filled either with soft tissue or, nearest to the skull bone, with ingrown bone (IB) which extends up to 4 mm from the skull bone. (Hematoxylin-toluidine blue acid fuchsin stain; magnification × 13.)

"islands" could be shown at distances of 3 to 4 mm from the adjacent bone, thus demonstrating the depth over which bone ingrowth could take place (Figure 21).

VI. VASCULARIZATION OF BONES

A. Introduction

The vascularization of all bones in the adult body is not yet unequivocally ascertained. Moreover, its changes during the remodeling of bones in adult life and the changes in vascularization in young, yet developing bones complicate the picture.

Most implantologists will have to deal with long bones (e.g., the femur with regard to arthroplasty), the jaws (in conjunction with tooth root or tooth replacements), and/or flat bones of the skull (the calvarium with regard to augmentations or defect restorations). For this reason, the description below has been limited to these bones. Moreover, the vascularization in the mature situation is described nearly exclusively.

Also in bone, as in other organs, the vascular system possesses an arterial (afferent) part and a venous (efferent) part which are interconnected by capillaries. Both the large afferent and efferent vessels branch into smaller vessels, the arterioles and venules which link up with the capillaries. The afferent vessels provide the bone with nutrition and oxygen and the efferent vessels remove newly formed blood cells from the bone marrow and/or carbon dioxide and other waste products.

B. Cranial Bones

The desmal bones of the calvarium, which is part of the neurocranium, enclose the brain. The periosteum is adjacent to the outer meninx (the dura mater).

From the dura mater, one or two nutrient arteries penetrate the developing skull bones (the osses frontale, parietale, and occipitale), i.e., their ossification center(s). These arteries ramify in the ossification centers. The ramifications run centrifugally towards the place where, at the time the bones are fully developed, the sutures are located.

An important constituent of the efferent system is the system of large sinuses which were described by Breschet in 1829. In the adult, these sinuses run close to the sutures of each bone in the diploë, i.e. the spongy bone between the external and internal table of the calvarian bones. These large veins leave via few wide foramina in the outer table and via many wide nutrient foramina in the inner table. In addition, a large number of small foramina can be observed near the sutures of the cranial bones as well externally as at the dural side. Fine arteries and veins pass these foramina which join the vascular system inside the cranial bones. Still smaller foramina can be observed which are scattered over the entire inner and outer surface of the cranial bones. Here, capillaries pass which run from the diploë to the external periosteum or the dura.

In the soft connective tissue of the sutures, veins can be found which interconnect the Breschet sinuses of one calvarian bone with those of another. Moreover, veins can be observed, running perpendicularly to those just mentioned, which interconnect the veins of the dura mater with the veins of the external periosteum.

It will be evident from the above that the arteries, veins, and capillaries in the flat bones of the skull are intensely interconnected with the circulatory system of the external periosteum and the dura mater.

Therefore, it will be easily understood that removal of the external periosteum, e.g., to apply a biomaterial to cranial bones for reasons of augmentation, does not lead to the death of these bones. The vascular supply from the dura and adjacent bones is sufficient for survival. Equally, when a part of a cranial bone is raised while retaining continuity of the external periosteum, the periosteal vascular system is sufficient for survival of the bone flap.

C. Alveolar Process

The alveolar process is that part of the jawbones that supports the teeth. In its most extensive form it consists of three parts:

1. The lamina dura, which envelopes the soft tissue of the periodontal ligament and the tooth root. It is a thin plate of lamellar and bundle bone.
2. The supporting alveolar bone. This bone is of the spongy type. The trabeculae transfer mechanical forces. These forces (e.g., masticatory forces) are transferred from the tooth to the lamina dura by the collagen fiber bundles of the periodontal ligament. Then the trabeculae transfer the forces to the cortex of the jawbones.
3. The third part of the bone of the alveolar process is that part of the cortical plates of the jawbones which is near the teeth. This bone is of the compact type.

The bony structures around the tooth socket are not always as elaborate as described above, e.g., the buccolingual aspects of the lamina dura of the incisors and canines in the upper and lower jaw are partly or fully fused with the cortical bone. In these cases an intermediate layer of spongy bone is missing. It is evident that in these cases the blood supply of the alveolar bone is provided by the vascularization of the mandibular and maxillar cortex, i.e., the blood vessels in the haversian canals and the canals of Volkmann.

Where interradicular and/or interdental septa exist, nutrient canals are present in them. These septa consist of the spongy supporting alveolar bone and the adjacent lamina(e) dura(e). The nutrient canals are called the canals of Zuckerkandl and Hirschfeld and not only lodge arteries and veins, but also lymph vessels and nerves. Branches of these blood vessels perforate the lamina dura and run in the periodontal ligament. In this way the osteocytes of the spongy bone and the sieve-like lamina dura are nourished.

D. Long Bones

Although different views on the vascularization of long bones exist, only one theory is reviewed here.

The afferent nutrient supply of long bones can be divided in three parts: the principal nutrient artery, the circulus articuli vasculosus, and the periosteal arterioles.

The principal nutrient artery passes through the cortex of the diaphysis and only branches in the medulla. There are two branches, the ascending and the descending medullary arteries. These subdivide into arterioles which not only supply the whole endosteal surface of the diaphysis, but also provide for the arterial supply of the entire cortex in the case the cortex is covered by a loosely attached periosteum. In the case a firmly attached periosteum is present (around sites of insertion of fasciae), only the inner ⅔ of the cortex is vascularized from medullary arteries.

The circulus articuli vasculosus divides in arteries to the metaphysis and arteries to the epiphysis. The metaphyseal and epiphyseal arteries pass through the metaphyseal and epiphyseal bone and divide into small arteries and arterioles which run in the spongy bone of the metaphysis and in the epiphysis. In addition, many anastomoses exist which interconnect the metaphyseal vascularization with the diaphyseal blood supply. This is an explanation why, in the case of rupture of the nutrient artery (e.g., in bone fracture), the diaphyseal system stays functional. Thus, nourishment of the diaphyseal cortex is maintained.

The periosteal arterioles enter the cortex around sites where fasciae penetrate the diaphysis. They supply vascularization to the outer ⅓ of the cortex in these locations. Anastomoses exist in the cortex interconnecting the periosteal supply with the arterioles of the medullary supply. By this fact the explanation is given why, after disturbance of the medullary vascularization, the cortical bone necrotizes all but that part of the cortex, the blood supply of which is provided by the periosteal arteries, i.e., that part of the cortex near insertion sites of fasciae.

The efferent vascular system also consists of three parts: the accompanying vein (vena nutritia comitans) of the principal nutrient artery together with some other large veins, the cortical venous channels, and the periosteal capillaries.

In the medulla of the diaphysis a venous sinus is located length-wise. This sinus receives blood from the innermost part of the compact cortical bone. It also receives blood which contains newly formed blood cells from the hematopoietic (red) bone marrow. The venous sinus drains into the nutrient vein and the other large veins which pass the cortex and empty in the veins of the surrounding soft tissue.

The large number of the cortical venous channels transport the blood from all cortical bone, except its innermost and outermost parts, to the veins of the soft tissue area around the bone.

Apart from the above-mentioned large veins, many capillaries emerge from the bone surface at sites where the periosteum is loosely attached to the bone. These blood vessels transport blood from the outermost part of the cortex to the vascularization of the soft tissue around the bone.

From the above, it will be clear that in the compact lamellar bone of the cortex of long bones the arterial blood flow, is in the direction from the medulla to the surrounding soft tissues. This is with the exception of a small number of large arteries which do not branch within the cortex and the periosteal arterioles which do divide. However, the sphere of influence of these vessels is limited to the areas where fasciae insert.

GENERAL REFERENCES

Bhaskar, S. N., *Orban's Oral Histology and Embryology*, 9th ed., C.V. Mosby, St. Louis, 1980.

Bloom, W. and Fawcett, D. W., *A Textbook of Histology*, 9th ed., W.B. Saunders, Philadelphia, 1968.

Bourne, G. H., *The Biochemistry and Physiology of Bone*, Vol. 2, 2nd ed., Academic Press, New York, 1972.

Brookes, M., *The Blood Supply of Bone: An Approach to Bone Biology*, Butterworths, London, 1971.

Ham, A. W., *Histology*, Pitman Medical, London, 1965.

Junqueira, L. C. and Carneiro, J., *Basic Histology*, Lange Medical, Los Altos, Calif., 1981.

Junqueira, L. C. and Carneiro, J., *Functionele Histologie*, Wetenschappelijke Uitgeverij Bunge, Utrecht, The Netherlands, 1981.

Ten Cate, A. R., *Oral Histology: Development, Structure, and Function*, C.V. Mosby, St. Louis, 1980.

Chapter 4

TISSUE INTERACTIONS TO BONE REPLACEMENT MATERIALS

Günther Heimke and Peter Griss

TABLE OF CONTENTS

I. INTRODUCTION

The materials used and presently considered as bone substitutes can be categorized as

Sufficiently biocompatible
Biocompatible
Bioinert
Biodegradable
Positively influencing bone formation

for which some definitions and examples are given in Table 1. Sufficiently biocompatible and biocompatible materials release something into the surrounding tissue, but of a kind and in an amount not severely preventing normal cell proliferation on or in some close vicinity of their surfaces. The bioinert materials do not influence the adjacent tissue biochemically. The materials which positively influence bone formation allow for some deposition of a mineral phase onto their surface if inserted into bony tissue and/or they allow for some kind of bonding to bony tissue.

Besides grouping the implant materials (replacing bony tissue) according to their degree or kind of compatibility, they could also be ordered according to the load situation of their intended applications:

- Heavy load bearing applications like the anchorage of joint endoprostheses, bridging gaps of long bones, also tooth root replacements.
- Low load-bearing applications like filling of holes after cleaning cysts or other defects that are still essentially surrounded by bone, mastoid fillings, replacement of bony walls like portions of the skull, paranasal sinuses, osseous walls of the external meatus, reconstruction of the ossicular chain.

In some of these applications, the implants must remain at full function for the rest of the life span of the patient. For others, at least theoretically, the implant materials could be resorbed if they are simultaneously replaced by newly formed bony tissue.

Because, from the present knowledge, there is no way of regenerating the natural function of the articulating surfaces of joints after they once have been fully destroyed, all joint endoprostheses must be permanently anchored. The same necessity exists for tooth root replacements. Therefore, in such cases, the implant material itself and the interface between the implant and the living tissue must maintain its load bearing and load transmitting abilities for many decades. Think of a tooth root replacement in a 10 year old child! For all of the low load bearing applications and also for bridging gaps of long bones one could think of, materials that serve their purpose temporarily but are being replaced more or less slowly and completely by well-organized living bone and the functional structures concerned: tooth, joint surfaces, etc.

In the studies of the tissue reactions to bioinert implants under load free or nearly load free conditions, it was found that all steps and processes of bone remodeling occur in the immediate vicinity of the implant surface that are known from fracture healing.[22] In these experiments it was noted that bone can grow above original cortical bone levels along a bioinert material such as Al_2O_3-ceramics as can be seen in Figure 1.[1] The same close contact of normally proliferating bony tissue is also found on those surfaces of Al_2O_3-ceramic implants into which the forces are transmitted perpendicular or at least nearly perpendicular. The load free and the pressure loaded interfaces are distinguished by the absence of relative movements between the implant surface and the bony tissue. Along the tangentially loaded interfaces (shearing forces) along which

Table 1
CLASSIFICATION OF IMPLANT MATERIALS FOR ORTHOPEDIC SURGERY

Class	Definition	Examples: typical, representative
Sufficiently compatible materials	Release of ions or other constituents influencing the surrounding tissue in an unfavorable, but still tolerable manner	Stainless steel, PMMA-cement
Compatible materials	Release of ions or other constituents not influencing the surrounding tissue appreciably	Co, Cr, Mo alloys, UHMW-polyethylene
Bioinert materials	No detectable release of ions or any other constituents	High purity Al_2O_3-ceramics
Biodegradable materials	Dissolve or otherwise disappear without causing any unfavorable reaction in bony tissue	Porous tricalcium phosphates
Materials positively influencing bone formation	Ability to bond to bony tissue in the sense of a glueing effect	Calcium phosphate ceramics, some calcium phosphates containing glasses

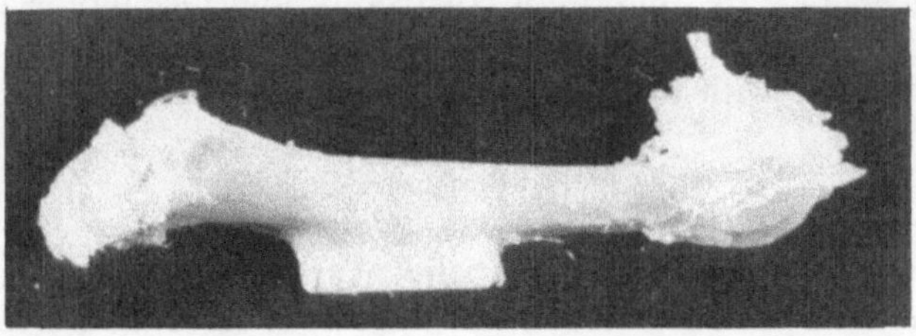

FIGURE 1. Biocompatibility test in a nearly loadfree position (midshaft of rat femur, approximately inserted along the neutral plane, if the femur is regarded as a cantilever).

any load change results in a relative movement, the formation of a soft tissue interlayer was observed.[23,27,66,67,34] Thus, only the pressure transmitting interfaces of a prosthetic device are available for its stable anchorage with all other surfaces hardly contributing to its fixation.

If, however, a true bonding between the implant and the bony tissue could be achieved, all of the surfaces could be available for implant fixation, thus allowing for more freedom in implant design and operational techniques. The formation of such bonds in the sense of a gluing effect along the surfaces of some calcium phosphate ceramics has increased chances of reaching this goal.

II. INTERFACE REMODELING ALONG BIOINERT IMPLANTS

As a prototype of an inert material high purity Al_2O_3-ceramics can be considered. This can be seen in a real load free situation by the tissue reaction along middle ear implants replacing the ossicular chain. Jahnke et al.[50,51] showed implants of this kind covered with a delicate middle ear mucosa within a few weeks after insertion. In considering this material for the application in a keratoprosthesis, Heimke and Polack[36] tested Al_2O_3-ceramic discs in the cornea and other soft tissue. A close adherence of fibroblasts to the surface of the implant material was found.[87]

As mentioned above, if implanted in hard tissue the load situation at the interface between bony tissue and any implant is of utmost importance for the judgment concerning the remodeling ability of the adjacent tissue.[39,42] This may be observed in soft tissue implants as a varying thickness of the fibrous tissue interlayer, its cellularity, and perhaps by the presence or lack of inflammatory response. The reaction of bony tissue adjacent to the implants may, in addition, be more subjected to load by its implantation mode, the implant shape, and the implant function. Deformation of the bone at the implant surface under load results from different kinds of stresses (strain, compressive, or shear) along the interface,[60] some of them producing relative movements and specific tissue reactions depending on the amount and the type of stress applied. Therefore, besides normally proliferating lamellar bone, a fibrous tissue layer or fibrous cartilage may be the answer to the local stress situation at different interface areas of the same implant.[23,27] This has caused considerable difficulty and invited much controversy about the qualitative morphological interpretation of interface tissue reactions for the rating of the biocompatibility. This was true until Heimke et al.[39,42] clearly stressed the importance of defining the main load condition at the interface of any histologic section presented and interpreted in papers to serve as a basis for comparative biocompatibility discussions. This requirement is far from generally being fulfilled and included in the planning and evaluation of biomaterials research. Besides other influences controlling the tissue reaction along the surfaces of an implant, e.g., implant toxicity (chemical dissolution or corrosion), interface biomechanics are an essential factor for tissue differentiation along implants. The detailed studies of these aspects become possible with the availability of highly inert materials like carbon or alumina.

A. Qualitative Tests in Bony Tissue

Nearly simultaneously, but mostly independent from each other, dense alumina implants were tested in the proximal femur in dogs by Boutin[7] and Hentrich et al.,[46] in the midshaft of the femur of rats by Griss et al.[21,30] (Figure 1), in the same area of dog femura by Predecki et al.,[88,89] and finally in the distal femur of rabbits by Cini et al.[12] All researchers report a rapid woven bone formation along smooth and profilated alumina surfaces within four to eight weeks in the different test animals. Mature lamellar bone in close contact to the alumina surface without visible soft tissue interlayer is seen within two to four months after implantation.[34] There is no significant further change of the bone/alumina implant interface beyond six months after implantation in experiments as just described. All authors mention the possibility of alumina to be integrated into bone without a soft tissue interlayer relatively soon after surgery (Figures 2 and 3). Hulbert et al.,[47,49] Klawitter et al.,[63] and Karbe et al.[58] have reported similar results using porous alumina implants in the midshaft of the dog femur. The direct bone to alumina contact may be very intimate. The calcified bone matrix follows microscopic surface undulations of the alumina so closely that a bony replica of crystalline surface of the ceramic can be seen in the scanning electron microscopy of Figure 4.[69] Osteocytes close to the surface of the alumina implant project their plasmatic dentrites up to the surface of the alumina implant (Figure 5).[30] This can be regarded as a very delicate proof of the bioinertness or biocompatibility of high density alumina.

Besides interface portions having this intimate bone to alumina contact, most implants show more or less extensive areas where alumina and bone are separated by fibrous tissue. This was especially pointed out by Köster et al.[66,67] and Blencke.[5] Griss et al.[21,24] also have reported chondroblastic metaplasias as well as fibrous tissue along alumina implants in the rat femur up to 70 weeks after implantation in addition to the just described intimate bone implant contact. Harms and Mäusle[34] were able to demonstrate a monocellular layer of macrophages between the alumina surface and neighboring soft or hard tissue.

These deviating results can, of course, be attributed to the different load patterns along different portions of the surface of the same implant or to different degrees of

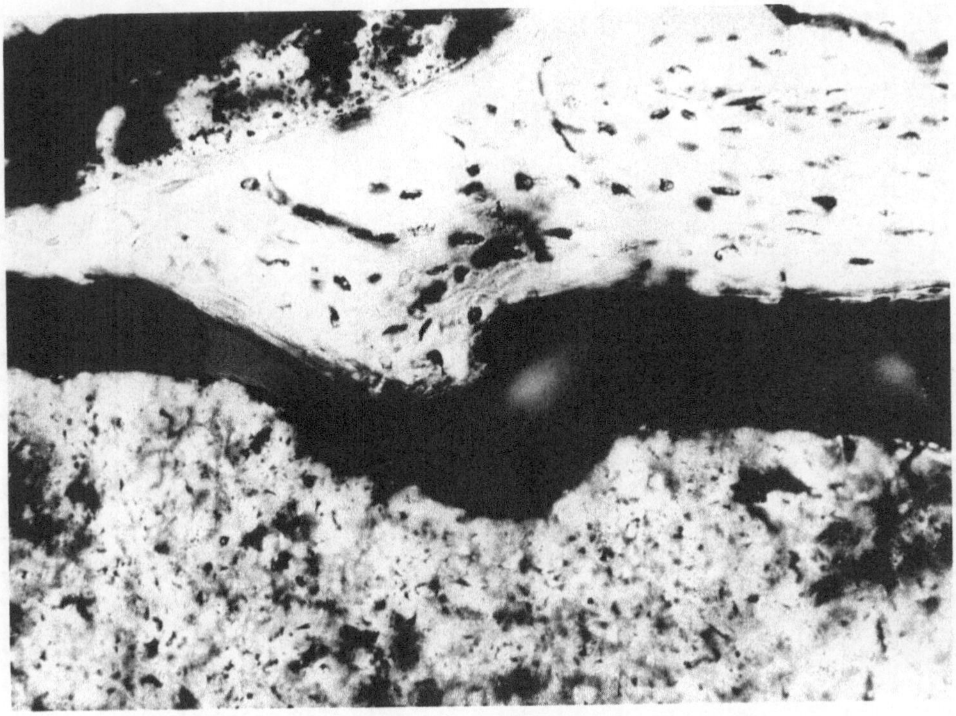

FIGURE 2. Al_2O_3-ceramic implant in bony tissue with areas of direct contact. Ceramic and tissue separated during preparation, but without soft tissue in between.

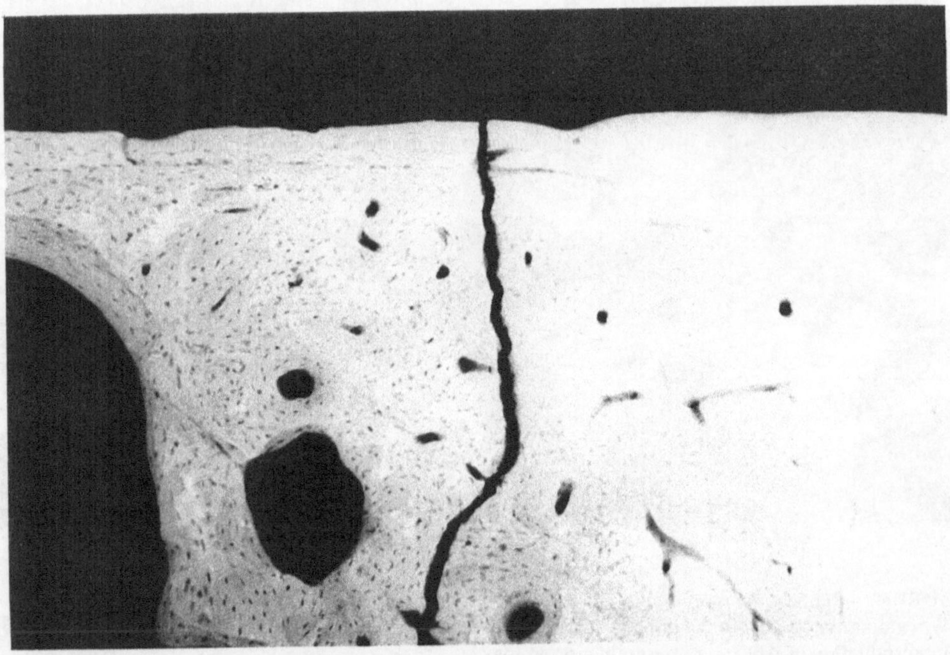

FIGURE 3. Microradiographic representation of the interface (top part) 8 months after implantation in rat femur, × 100 showing smooth and solid lamellar bone structure along the surface.

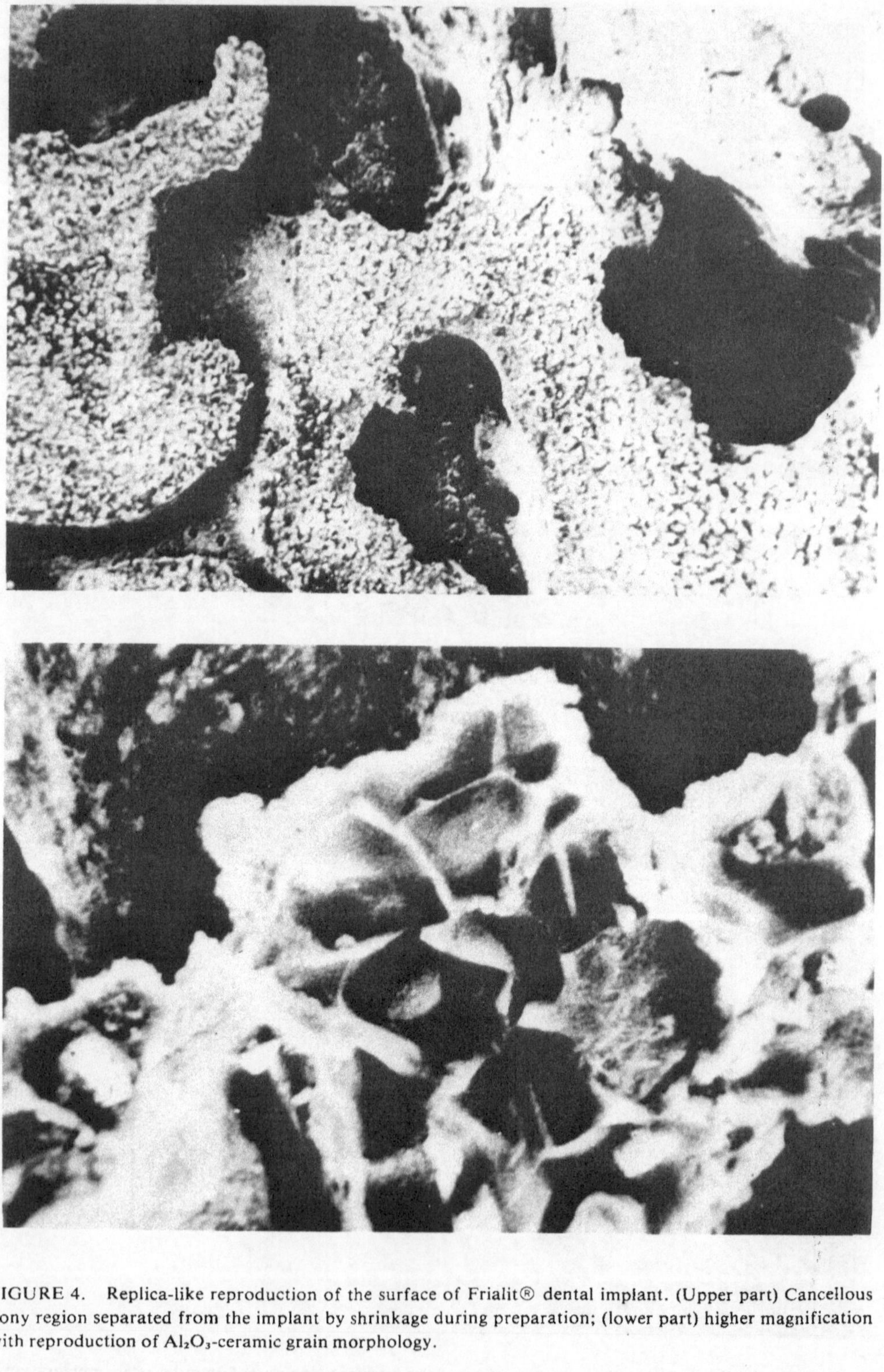

FIGURE 4. Replica-like reproduction of the surface of Frialit® dental implant. (Upper part) Cancellous bony region separated from the implant by shrinkage during preparation; (lower part) higher magnification with reproduction of Al_2O_3-ceramic grain morphology.

stabilization of different implants. Therefore, in order to achieve a reliable mechanical stabilization, an exact surgical technique is necessary for undisturbed bone maturation at the implant interface. Any excess of relative movements at the interface reduces the

FIGURE 5. Osteocyte projecting its plasmatic dendrites towards the surface of the alumina implant.

potential of the initially formed fiber callus to transform into woven and lamellar bone. The final outcome of every interface remodeling after implanting a biomaterial into bone largely depends on local mechanical factors.[39,42] This is true for alumina as well as for inert glasses, carbon, metals, and alloys, or plastics. In the midshaft of the femur, a mechanically well-defined load situation is only approximated close to the center plain, i.e., the neutral plain of the femoral tube in lateromedical projection. Comparative biocompatibility studies in bone should therefore be restricted to this central marrow area of the implantation site (Figure 6). Only when close to this area are relative movements of the implant against the surrounding tissue negligibly low. Experience shows that direct bone implant contact is only observed along this portion of the implantation site.

Cancellous bone implantation sites as the trochanter or the epiphysis of femur and tibia present difficult biomechanical load patterns which do not allow for an exact prediction of the biomechanical situation along the different surfaces of an implant. Mechanically poorly defined implantation sites explain most of the diverging descriptions of a biomaterial's compatibility based on the morphological interpretation of the tissue interface. Therefore, the midshaft implantation technique as first described by Beckham et al.[3] can be recommended as a standard biocompatibility test in bone approximating nonloaded conditions.

As the authors mentioned above, these mechanical influences on biocompatibility tests can be seen clearly and are undisturbed by other influences in experiments with chemically inert implant materials such as high purity alumina. All other implant materials (in a positive sense: bioglass,[43,6] calcium phosphates;[17,54,96] in a negative sense: many plastics, glasses, and a number of alloys) more or less influence and sometimes even obscure this mechanical interface maturation by ion release (favorable or toxic) from the implant material. The qualitative morphological interpretation of "one selected interface area" around any implant in bone is therefore by no means sufficient for biocompatibility predictions.

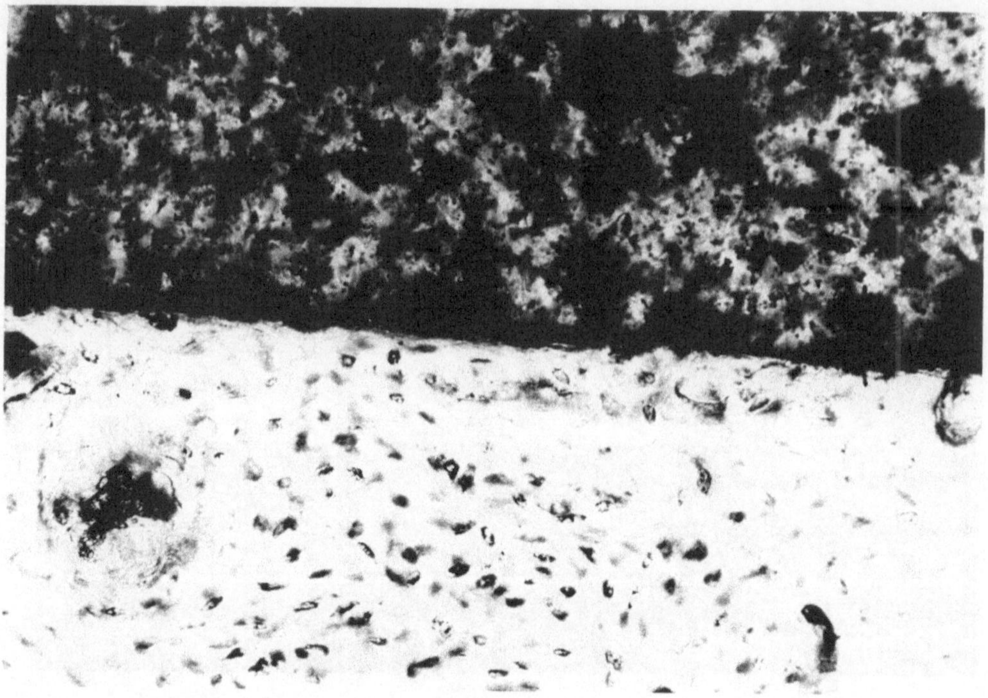

FIGURE 6. Histological cross-section of an implant located along the "neutral plane" as mentioned in Figure 1.

B. Quantitative Biocompatibility Tests

The quantitative morphological biocompatibility testing of Al_2O_3-ceramic was done in comparison to AISi 316 stainless steel using the rat femur implant model as described above.[21,23,24,30]

Characteristic differences can be seen by differential cell counting within the fiber-callus layer at the interface 1 to 3 months after implantation of alumina and stainless steel. When fibroblasts and histiocytes, fibrocytes, and round cells are counted separately within the fibrous capsule one month after implantation, immature fibroblasts predominate around alumina. The number of fibroblasts decreases continuously until three months after surgery while the number of mature fibrocytes is increasing proportionally. This can be regarded as a physiological healing-in-process. The number of fibroblasts around stainless steel is considerably lower within the first two months and decreases only in the third month after surgery. Fibrocytes are observed in a slowly rising number within the examination period. Round cells are found to be slightly, but not significantly higher around steel than around alumina. The difference between fibroblasts and fibrocytes is statistically significant only after one month. This may be interpreted as a more rapid decrease of the number of fibroblasts around steel.

Studies using ^{3}H-thymidine as a DNS-synthesis marker show S-phase cells in close contact to alumina as well as to stainless steel. Cell counting of ^{3}H-thymidine-marked S-phase cells (^{3}H-index) shows a synchronized behavior around both materials without significant difference. Differential cell counting and ^{3}H-index determination, therefore, do not reveal striking differences between stainless steel and alumina in these short-term biocompatibility tests. This corresponds well with qualitative morphological studies in which a relatively low cellulated thin fibrous tissue capsule is formed around alumina[48] as well as around stainless steel[14] after implantation in soft tissue (s.c. or i.m.).

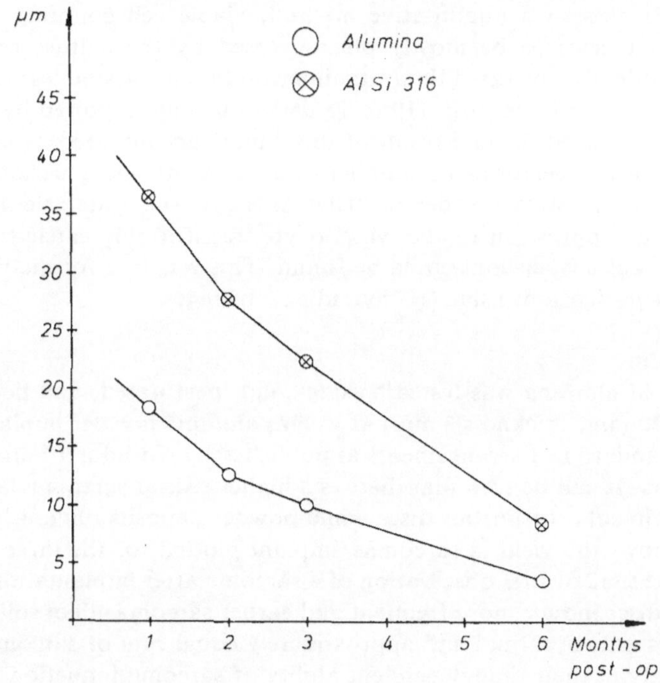

FIGURE 7. Thickness of layer of S-phase cells along alumina and AlSi 316 stainless steel as a function of time.

Comparative thickness measurements of the fiber-callus interface were performed on ground sections of the implantation site without removing the alumina and stainless steel implants after implantation periods of zero to six months. The results of one such study are shown in Figure 7. The thickness around alumina is significantly lower at one to three months. It decreases rapidly between three to six months after implantation. Around stainless steel the soft tissue interlayer is significantly thicker one to three months after implantation. Its values later come close to those measured around alumina. This "dynamic" difference in relation to time may be interpreted as a quicker and more undisturbed bone formation out of immature fiber callus around alumina.

C. Tissue Culture Experiments

Tissue culture studies with alumina had been reported by Heimke et al.,[38] Kawahara et al.,[61] and Plenk.[86] More recently, Harms and Mäusle[34] have also reported tissue culture studies with alumina. Heimke et al.[38] compared the behavior of embryonic fibroblast cultures (mouse) after addition of alumina and Protasul 10® test samples. They found a remarkable difference in cytotoxicity, with alumina being tolerated almost without reaction by the cultivated fibroblast. Kawahara[61] discussed the importance of the polarized alumina surface in wet milieu. This allows polarized organic molecules in the body environment to attach to alumina by Coulomb rather than van der Waal forces and to cover alumina with the body's own proteins, mucopolysaccharides, etc., as had been pointed out by Heimke et al.[37] In addition, this mechanism promotes cell adhesion to alumina surfaces as a basic positive principle of tissue biocompatibility. Many plastics, for example, do not show cell adhesion. In numerous cell culture experiments using different cell-lines, Kawahara[61] demonstrated the excellent capability of cells to adhere to single crystal or polycrystalline alumina.

Alumina powder (medium grain size 5 μm) was tested by Plenk in human fibrocyte cultures using Plenk's own quantitative method, viable cell counting. That author found alumina particles to be mostly phagocytosed by the culture cells. Alumina seemed to be a little less tolerated than titanium, tantalum, or stainless steel particles. The latter, however, were too large (10 to 15 μm) to be phagocytosed by culture cells. This may have influenced the end result of this study and obscured its interpretation.

Harms and Mausle[34] performed a number of experiments using periotoneal macrophage cultures and alumina powder particles as a test substrate. Neither a specific cytotoxicity nor a suppression of the lymphocyte transforming capacity of alumina-exposed macrophage suspensions could be found. This was true for qualitative as well as quantitative investigations using ^{3}H-thymidine labeling.

D. Sarcoma Studies

Sarcogenicity of alumina was tested[28] using solid, perforated, and porous alumina discs (diameter 20 mm, thickness 3 mm) as well as alumina powder implanted s.c. into rats following standard test arrangements as published by Nothdurft[80] and Ott.[83] Table 2 gives the results. It can be seen that there is a higher risk of sarcoma development in perforated than in solid or porous discs while powder deposits obviously have a zero risk. Figure 8 shows the yield in sarcomas/implant plotted for the three differing implants against the time of first observation of a sarcoma after implantation. This shows that perforated discs induce more frequent and earlier sarcomas than solid discs. Solid and porous discs induce a final and approximately equal rate of sarcomas, but these differ significantly in their time-dependent ability of sarcoma induction. Their different abilities of sarcoma induction can easily be explained on a morphological basis. Hulbert et al.[48] (1972) have described thicker fibrous tissue encapsulations around solid discs than around porous discs. Macroporous as well as perforated discs, however, show the same thickness of encapsulation as solid discs, though more than solid discs due to their larger surface. As sarcoma induction is clearly a function of the amount of scar tissue present, this phenomenon is easily explained. The most important clinical issue of this graph (Figure 8), however, is the fact that macroprofiliated implants may be more sensitive to sarcoma induction than smoothly surfaced or porous implants.

Whether such foreign body sarcoma studies in animals have a clinical relevance is still a matter of controversy. Foreign body sarcomas could be induced with a wide variety of materials independent of their chemical composition.[83] Nothdurft[80] could clearly relate this phenomenon to surface and scar tissue amount rather than to any chemical or other sarcoma induction mechanism as proposed by Oppenheimer et al.[82] The sarcoma rate around alumina test specimens is aligned with those reported on other compounds and lower than those found with many plastic materials.[83] Probably the most important clinical relevance of this study was the zero percent sarcoma rate around powdered alumina as a clear indication for its high compatibility and low fibrogenic potential.

E. Functionally Loaded Implants

Most bone replacements do involve temporary or permanent load bearing situations. The importance of the local load pattern for the tissue response at the site concerned has been explained.[39,42] The interface histology along loosened inert implants resembles that of pseudoarthrosis tissue. Therefore, this was called "interface pseudoarthrosis".[22,23] Krempien et al.[68] have given a detailed description of this situation. Along purely pressure-transmitting interfaces the bone remodeling is essentially identical as along load free areas if the implant is rigidly fixed.[22,25,29] This was further confirmed with tooth root replacements carrying design features particularly intended for stabilization by pressure-transmitting interfaces.[93,94,9,10]

Table 2
SUMMARY OF THE RESULTS OF A SARCOGENICITY TEST OF HIGH PURITY ALUMINA

Kind of implant	No. animals	No. implants	Sarcoma/animal	Sarcoma/ implant
Solid discs	33	264	1.4	0.17
Perforated discs	27	216	2.0	0.25
Porous discs	21	84	0.8	0.18
Powder	47	376	0	0
Control	22	0	0	0

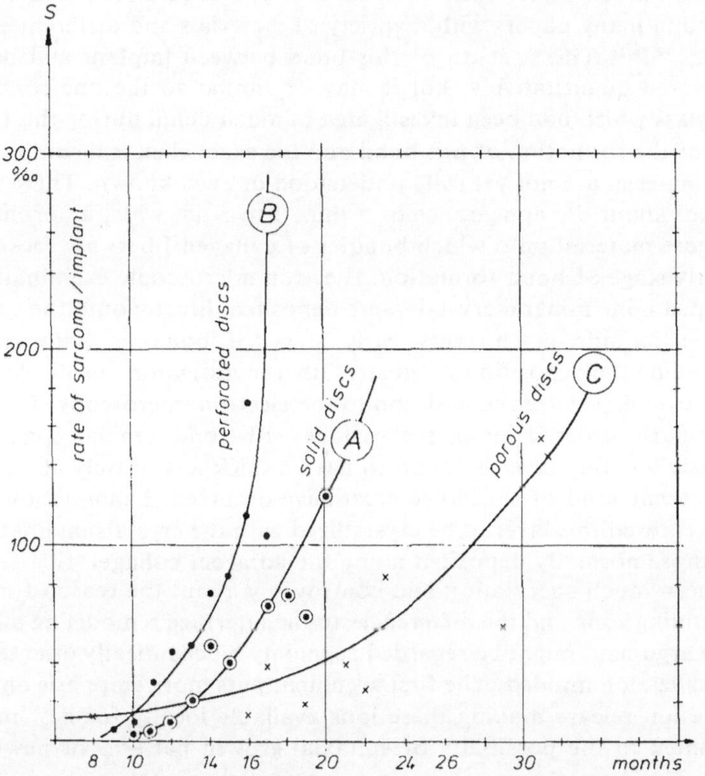

FIGURE 8. Sarcomas per implant (in 0/00) as a function of time after implantation of three types of ceramic discs.

III. INTERFACE REACTIONS ALONG CALCIUM PHOSPHATE CERAMICS

The interface remodeling associated with implantation of dense and nonresorbable calcium phosphate ceramics in bony tissue can also be characterized as representing normal bone healing processes.[52,15,59,16,79,81,97] Along dense as well as porous materials, normally proliferating bony tissue is usually found deposited directly on the surfaces without the presence of an intervening fibrous tissue layer. In several studies, using electron microprobe point counts, ratemeter scans, pulse images, or less quantitative techniques such as microradiography and energy-dispersive analysis normal calcification processes could be detected in the immediate vicinity of the implants with Ca/P-ratios increasing from 1.50 at 1 month to 1.62 at 6 months, and simultaneously increasing calcium and phosphorous concentrations reaching the level of the surrounding bone toward the end of this period.[2,52,91] Some differences, however, are reported for more interior pores of thoroughly porous bodies along which the described processes obviously become retarded.[26,35,74,77,91,19,95]

A. Bond Formation

The most striking feature of calcium phosphate ceramics that distinguishes these materials from the bioinert implant materials is their ability to become directly bonded to bone. This was first observed by Driskell et al.[18] who reported that fractures propagated through the bone-ceramic interface from which observation the conclusion was drawn that a chemical bond would exist between the bony tissue and the ceramic. Since then, the formation of a direct bond between this type of ceramic and living tissue has been confirmed in many papers with a variety of materials and different experimental arrangements.[52,59,15,16] The strength of this bond between implant and bone has not yet been measured quantitatively, but it may be similar to the one formed between bone and bioglass which had been investigated in much detail during the 1970s.[44,45,8,97]

The details of the formation of this bond and the exact chemical composition of the "cementing" material are not yet fully understood or even known. There is, however, some agreement about the appearance of a thin, 3 to 5 μm wide, amorphous layer of an electron-dense material onto which bundles of collagen fibers are loosely arranged in the very early stage of bond formation. Electron microscopic examination revealed that plate-shaped bone mineral crystals were deposited directly onto the surface of the hydroxylapatite ceramic in the very early stage of bonding, with some investigators[53,57,81] finding the deposition irregular. Other investigators, using electron microprobe and electron diffraction in addition to the electron microscopy, found the crystals perpendicularly ordered along the surface with some regular spacing. In later stages, this basic bonding layer is found to have a thickness of only about 0.05 to 0.2 μm; therefore, some kind of shrinkage must have occurred. Examination of undecalcified sections showed this layer to be crystallized with the crystals interacting or interlocking with those normally deposited along the adjacent collagen fibers.[52,57,53,62,81,16]

There has been much speculation and controversy about the reason for the formation of this bonding zone and the differences to the interface remodeling along bioinert implants. One argument might be regarded as mainly biochemically oriented, the other as more mineralization minded. The first argument puts more emphasis on the calcium and phosphate ion release making these ions available locally for new precipitation. The other pointed to the possibility of epitaxial growth patterns or newly deposited crystals.

B. The Solution of Calcium and Phosphate Ions

The paths of released calcium and phosphate ions from resorbable and nearly nondegradable materials have been traced with different methods such as serum and urine calcium and phosphate levels, and in different animal models without any abnormal results. In addition, no particulars or pathological reactions were found in detailed histological studies on the major organs including very detailed kidney histology.[76,77,11,71,19,95] These results lead to the conclusion that the calcium and phosphate ions dissolved from these ceramics are integrated into the usual regulatory system of the body and are handled (deposited or removed) normally. A very detailed study of the fate of calcium released from a bioresorbable tricalcium phosphate implanted in the calvaria of rabbits fully confirmed this conclusion. The liver, skin, brain, heart, kidney, lung, and intestine showed no abnormal content of the labeled calcium. Only the tibia was found to contain measurable amounts of Ca^{45}, thus showing some accumulation only in those areas in which the metabolism is much slower.[40]

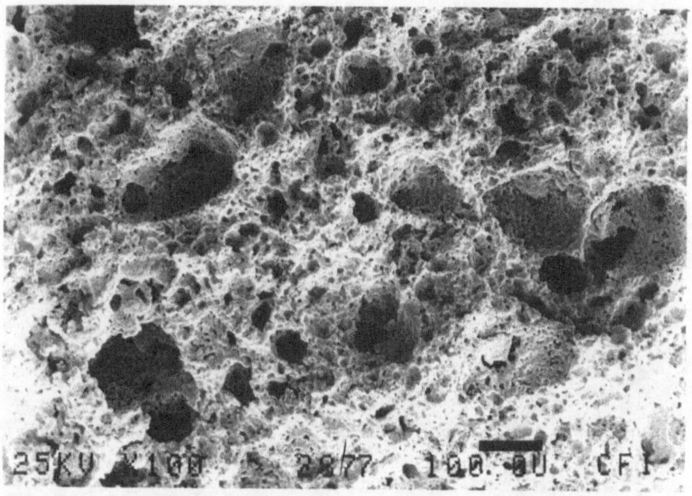

FIGURE 9. Tricalcium phosphate ceramic with pores of about 150 to 200 μm opening (macropores) surrounded by walls containing micropores.

C. The Resorption or Disintegration of Degradable Ceramics

There are no calcium phosphate ceramics that can be regarded as completely nonsoluble. Thus, the concept of bioresorption is one of degree and definition. But there are also other processes of degradation besides those that are solution-mediated.

The factors determining the resorption rate of these ceramics are still not fully understood. The experimental results are sometimes confusing. Two completely different degradation processes are considered: a more chemically, solution minded one and a more structurally oriented point of view. Of course, a solution-controlled process must be related to the surface that is exposed to the solvent, as was found in comparing the relative dissolution rates in buffered lactic acid of human dental enamel and a dense hydroxylapatite ceramic.[41,70,53] They became identical if the normalized surface areas were compared. Contrary to expectations, the material more closely resembling the body's own hard tissue component, the hydroxylapatite, was found to dissolve much more slowly than the tricalcium phosphates not naturally occurring in bone if similar ceramic structures and similar degrees of purity were used.[55,56] This was shown to be true in vitro with dense, pure ceramics in buffered lactate acid and buffered edetate basic media as well as in vivo with porous implants even after prolonged periods of implantation (more than one year).[26,35,78] Other in vivo studies using differently composed (tri- and tetra-), but structurally similar calcium phosphates also showed the composition dependence of the resorption rate.[65,31]

On the other hand, structural factors alone can determine the resorption rate and completely override chemical considerations. It is not yet clear if it is the surface area alone that controls these processes. The following approach to the problem seems to be most helpful[31] and obtains more and more experimental support.[97] Most of the porous calcium phosphate ceramics contain two major groups of pores: one with openings in the order of 100 to 500 μm (Figure 9) for allowing the ingrowth of tissue including the supplying blood vessels[63] and the other contained in the walls surrounding the openings being only less than a few microns wide (Figure 10). Looking at these walls from the manufacturing point of view, they can be regarded as being mainly composed of the original calcium phosphate particles between which bridges have formed during sintering (neck formation). In this picture the "pore size" is mainly controlled by the

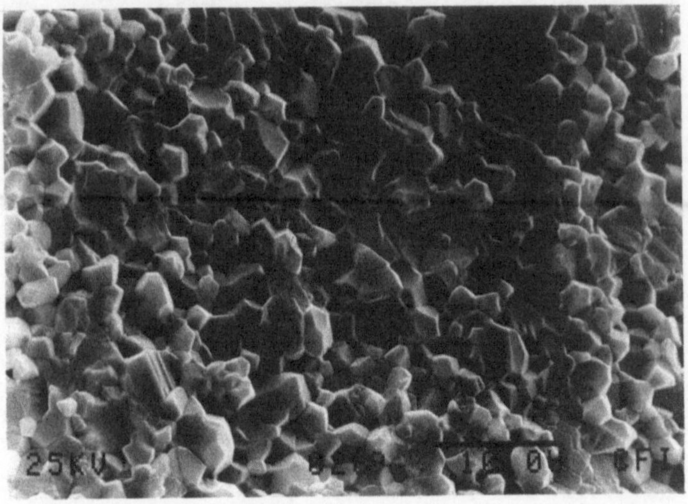

FIGURE 10. The same ceramic as in Figure 9 with higher magnification showing the microporosity in the walls.

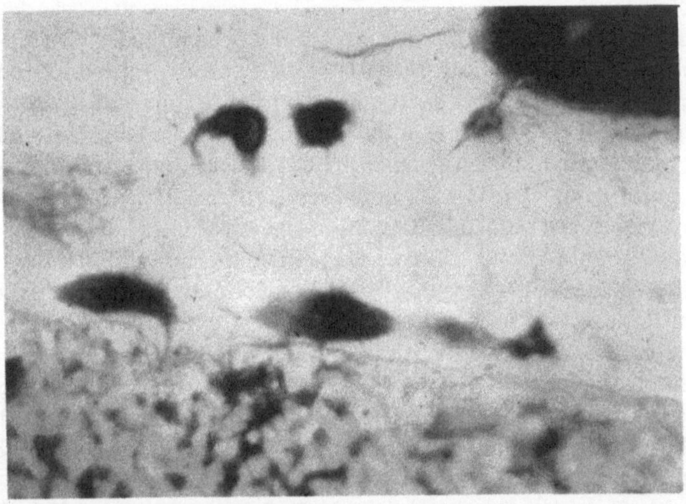

FIGURE 11. Osteocyte in close vicinity of a calcium phosphate implant with plasmatic dendrites extending into micropores.

size of the particles, and to a much lesser degree by the dimensions of the "necks". This "neck size", however, is strongly dependent on the "activity" of the starting powder and the sintering conditions.

The biodegradation can be regarded as starting at these "necks". If all these "tiny" necks which hold a particular particle are dissolved away, this particle will start floating freely and therefore can be phagocytosed. Experimental evidence in support of this hypothesis has been given by several authors.[4,11,84,19,31] The additional observation that osteocyte dendrites reach directly to the implant surface (Figure 11) and further into the micropores[97] can be interpreted in two ways: either these dendrites only follow the concentration gradient of whatever component toward the interface or they play an active role in the solution process of the neck, for example, by absorbing parts of the dissolved component. The fact that such dendrites have also been observed aiming at the surface of the bioinert Al_2O_3-ceramic implants[24] (as shown in Figure 5) seems to exclude the calcium or phosphate ion concentration gradient as a sole stimulus for this activity.

The details of the stimulation of the cell activities around these materials are not yet understood. The possible involvement of the immune system must be considered as done by Klein et al.[64] as well as the influences of electrochemical redox processes as have recently been summarized, in view of cardiovascular applications, by Wolf.[98]

D. The Osteogenicity of Calcium Phosphate Ceramics

There has also been some controversy and speculation about the osteogenicity of calcium phosphate ceramics.[20,33] All studies, however, in which these materials were examined in true bone induction models, produced negative results in quite the same manner as was found with the bioglasses mentioned earlier.[85,13,82]

In these experiments, the ceramics concerned had been partially combined with autogenous bone marrow. The results indicate that these ceramics can be advantageously used as carriers for osteogenic marrows.[75]

E. Defect Fillings

Many of the studies previously mentioned as well as many others[97] compared the regeneration of artificial defects after filling with autogenous or other graft materials, calcium phosphate ceramics, various mixtures of them, or empty holes, using different implantation sites and various kinds of animals.[73,74,95,72] In an attempt to quantitatively measure the time required to achieve bridging lesions of long bone, a porous tricalcium phosphate ceramic in granular form was compared with autogenous cancellous bone and with a 50/50 mixture of the two. The times required for bridging were 10 to 16 weeks, 4 to 6 weeks, and 6 weeks, respectively, confirming the superiority of the autogenous material already consistently found in attempts to histologically determine the healing rates. Using porous forms of calcium phosphate ceramics control defects that had been left empty have consistently been found to repair more rapidly than the filled ones. This result is the more important if the properties of the blood clot that cause it to shrink and retract from the inner walls of the holes are considered. This loss of contact of the clot's fibrin filaments to the bony environment will be greatly prevented if the defect is more or less completely filled with porous ceramics thus approaching the requirements of fast bone regeneration as established and described in detail by Schulte.[92] If dense hydroxylapatite ceramics in particulate form are applied to such surgically created defects, the healing rates have been found to be similar to those of the empty controls.[52]

IV. FINAL REMARKS

From the standpoint of hard tissue response to implant materials the calcium phosphate ceramics are probably the most compatible materials presently known. Thus, much research work is presently devoted to this material, attracting many researchers because of its "attraction" to bony tissue. Of course, a considerable portion of these activities are aiming at solving more application-oriented problems such as reliably coating such materials onto mechanically, sufficiently strong substrates and, at the same time, avoiding all leaching off of such necessarily thin films. But even if all these efforts do not lead to solutions allowing too much deviation from the purely biomechanical requirements for implant design, some additional instruction for a more advanced understanding of interface tissue reactions can be expected.

REFERENCES

1. von Andrian-Werburg, H., Griss, P., Krempien, B., and Heimke, G., Klinische Problematik und experimentelle morphologische Befunde bei der Werwendung keramischer Werkstoffe in der Orthopädie und Unfallchirurgie, *Z. Orthop.*, 111, 577, 1973.
2. Aoki, H., Kato, K., Ogiso, M., and Tabata, T., Studies on the application of apatite to dental materials, *J. Dent. Eng. (Jpn)*, 18, 86, 1977.
3. Beckham, C. A., Greenlee, T. K., and Crebo, A. R., Bone formation at a ceramic implant interface, *Calcif. Tissue Res.*, 8, 165, 1971.
4. Bhaskar, S. N., Brady, J. M., Getter, L., Grower, M. P., and Driskell, T., Biodegradable ceramic implants in bone, *Oral Surg.*, 32, 336, 1971.
5. Blencke, B.-A., Derzeitiger Stand der klinischen Anwendung von Keramiken fur den Knochen- und Gelenkersatz, *Orthopäde*, 7, 43, 1978.
6. Blencke, B.-A., Brömer, H., and Pfeil, E., Rasterelektronenmikroskopische Untersuchungen der Reaktion des Knochens auf glaskeramische Implantate, *Z. Orthop.*, 112, 978, 1974.
7. Boutin, P., Arthroplastie totale de la hanche par prothèse en alumine frittée. Etude experimentale et premières applications cliniques, *Revue Chir. Orthop.*, 58, 229, 1972.
8. Bunte, M., Strunz, V., Gross, U. M., Brömer, H., and Deutscher, K., Vergleichende Untersuchungen über die Haftung verschiedener Implantatmaterialien im Knochen, *Dtsch. Zahnatrztl. Z.*, 32, 825, 1977.
9. Büsing, C. M., Heimke, G., Schulte, W., and Linn, W., Gibt es ein Lastschatten-Phänomen? Biomechanische Aspekte bei Zahnimplantaten aus Al_2O_3-Keramik, *Pathologe*, 1, 56, 1979.
10. Büsing, C. M., Schulte, W., d'Hoedt, B., and Heimke, G., Histological results with biomechanically shaped implants with Al_2O_3-ceramic, in *Dental Implants*, Heimke, G., Ed., Carl Hanser Verlag, Munich, 1980, 21.
11. Cameron, H. U., MacNab, I., and Pilliar, R. M., Evaluation of a biodegradable ceramic, *J. Biomed. Mater. Res.*, 11, 179, 1977.
12. Cini, L., Sandrolini, S., Paltrinieri, M., Trentani, C., Pizzoferrato, A., and Zolezzi, C., Impianti senza cemento di artroprotesi d'anca con riporto superficiale de allumina, *Chir. Organi Mov.*, 62, 1, 1975.
13. Boyne, T. J., Fremming, B. D., Walsh, R., and Jarcho, M., Evaluation of a ceramic hydroxyapatite in femoral defects, *J. Dent. Res.*, 57A, 108, 1978.
14. Contzen, H., Straumann, F., and Paschke, E., Eds., *Grundlagen der Alloplastik mit Metallen und Kunststoffen*, Georg Thieme Verlag, Stuttgart, 1967.
15. Denissen, H. W. and de Groot, K., Immediate dental root implants from synthetic dense calcium hydroxylapatite, *J. Prosthet. Dent.*, 42, 551, 1979.
16. Denissen, H. W., de Groot, K., Kakkes, P., van den Hooff, A., and Klopper, P. J., Animal and human studies of sintered hydroxyapatite as a material for tooth root implants, 1st World Biomat. Congr. Baden, Austria, 3.8.1. (Abstr.), 1980.
17. Driskell, T. D., O'Hara, M. J., Sheets, H. D., Greene, G. W., and Natiella, J. R., Development of ceramic and ceramic composite devices for maxillofacial applications, *J. Biomed. Mater. Res. Symp.*, 2, 345, 1972.
18. Driskell, T. D., Hassler, C. R., Tennery, V. J., McCoy, I. R., and Clarke, W. J., Calcium phosphate resorbable ceramics: a potential alternative to bone grafting, *J. Dent. Res.*, 52, 123, 1973a.
19. Ferraro, J. W., Experimental evaluation of ceramic calcium phosphate as a substitute for bone grafts, *Plast. Reconstruct. Surg.*, 63, 634, 1979.
20. Getter, L., Bhaskar, S. N., Cutright, D. E., Perez, B., Brady, J. M., Driskell, T. D., and O'Hara, M. J., Three biodegradable calcium phosphate slurry implants in bone, *J. Oral. Surg.*, 30, 263, 1972.
21. Griss, P., von Andrian-Werburg, H., Krempien, B., and Heimke, G., Biological activity and histocompatibility of dense Al_2O_3/MgO ceramic implants in rats, *J. Biomed. Mater. Res. Symp.*, 4, 453, 1973.
22. Griss, P., Heimke, G., Krempien, B., Silber, R., Haehner, K., and Merkle, B., Erste Erfahrungen mit der Keramik-Metallverbundprothese, *Med. Orthop. Technik*, 95, 159, 1975.
23. Griss, P., Heimke, G., Krempien, B., Hartung, H. J., Reipa, S., Lauterbach, H. J., and Diem, Th., Experimental ceramic joint replacement, in *Biocompatibility of Implant Materials*, Williams, D. F., Ed., Sector Publishing, London, 1976, 70.
24. Griss, P., Heimke, G., and Krempien, B., Experimentelle Untersuchungen über die Verträglichkeit von Keramikwerkstoffen, *Med. Orthop. Technik*, 96, 105, 1976.
25. Griss, P., Die Aluminiumoxidkeramik — ein neuer Werkstoff für Endoprothesen. Ergebnisse der experimentellen Prüfung und erster klinischer Anwendungen, Thesis, University of Heidelberg, 1976.
26. Driskell, T. D., Hassler, C. R., Tennery, V. J., McCoy, L. R., and Clarke, W. J., Calcium phosphate resorbable ceramics: a potential alternative to bone grafting, *J. Dent. Res.*, 52, 123, 1973b.

27. Griss, P., Silber, R., Merkle, B., Haehner, K., Heimke, G., and Krempien, B., Biomechanically induced tissue reactions after Al_2O_3-ceramic hip joint replacement. Experimental and early clinical results, *J. Biomed. Mater. Res. Symp.*, 7, 519, 1976.
28. Griss, P., Werner, E., Buchinger, R., and Büsing, C.-M., Zur Frage der unspezifischen Sarkomentstehung um Al_2O_3-keramische Implantate, *Arch. Orthop. Unfall Chir.*, 90, 29, 1977.
29. Griss, P., Werner, E., Heimke, G., and Raute-Kreinsen, U., Vergleichende experimentelle Untersuchungen an Al_2O_3-Keramik und mit modifiziertem Bioglas (L.L.Hench) beschichteter Al_2O_3-Keramik. II. Ergebnisse mit belasteten Implantaten, *Arch. Orthop. Unfall Chir.*, 92, 199, 1978.
30. Griss, P., Werner, E., and Heimke, G., Alumina ceramic, bioglass, and silicon nitride: a comparative biocompatibility study, in *Mechanical Properties of Biomaterials*, Hastings, G. W. and Williams, D. F., Eds., John Wiley & Sons, New York, 1980, 217.
31. de Groot, K., Bioceramics consisting of calcium phosphate salts, *Biomaterials*, 1, 47, 1980.
32. Gross, U. M., Pickartz, H., Taube, K.-G., and Strunz, V., Tissue reactions during the incorporation of glass ceramic material, in *Evaluation of Biomaterials*, Winter, G. D., Leray, J. L., and deGroot, K., Eds., John Wiley & Sons, New York, 1980, 413.
33. Grower, M. F., Horan, M., Miller, R., and Getter, L., Bone inductive potential of biodegradable ceramic in millipore filter chambers, *J. Dent. Res.*, 52, 160, 1973.
34. Harms, J. and Mäusle, E., Tissue reaction to ceramic implant material, *J. Biomed. Mater. Res.*, 13, 67, 1979.
35. Hassler, C. R., McCoy, L. G., and Rotaru, J. H., Long term implants of solid tricalcium phosphate, *Proc. 27th Ann. Conf. Eng. Med. Bio.*, 16, 488, 1974.
36. Heimke, G. and Polack, F. M., Keramische Keratoprothesen, *Ber. Dtsch. Opthamol. Gesellsch.*, 75, 28, 1977.
37. Heimke, G., Griss, P., and Krempien, B., Ceramic endoprostheses, *Powder Metall. Int.*, 8, 77, 1976.
38. Heimke, G., Griss, P., Jentschura, G., Krempien, B., Gugel, E., Petzenhauser, I., and Hennicke, H. W., Keramische und keramischbeschichtete Knochenersatzwerkstoffe, *BMFT-Forschung.*, T77-70, 1977.
39. Heimke, G., Griss, P., Jentschura, G., and Werner, E., Die Aussagefähigkeit histologischer Befunde zur Beurteilung von Knochenersatzwerkstoffen, *Arch. Orthop. Traumat. Surg.*, 91, 267, 1978.
40. Hassler, C. R., McCoy, L. G., and Clarke, L. C., Studies on the degradability of large tricalcium phosphate segments, *Proc. 2nd Ann. Meet. Soc. Biomater.*, 1976, 88.
41. Huffman, E. O., Cate, W. E., Deming, M. E., and Elmore, K. L., Rates of solution of calcium phosphates in phosphoric acid solutions, *Agric. Food Chem.*, 5, 266, 1957.
42. Heimke, G., Griss, P., Werner, E., and Jentschura, G., The effects of mechanical factors on biocompatibility tests, *J. Biomed. Eng.*, 3, 209, 1981.
43. Hench, L. L., Splinter, R. J., Allen, W. C., and Greenlee, T. K., Bonding mechanisms at the interface of ceramic prosthetic material, *J. Biomed. Mater. Res. Symp.*, 2, 117, 1971.
44. Hench, L. L., Development of a new biomaterial prosthetic device, in *Orthopedic Mechanics Procedures and Devices*, Ghista, D. H. and Roaf, R., Eds., Academic Press, New York, 1978, 287.
45. Hench, L. L., Petty, R. W., and Piotrowski, G., An investigation of bonding mechanisms at the interface of a prosthetic material, Report 9, U.S. Army R. and D. Command, Contract No. DADA 17-70-C-0001, U.S. Government Printing Office, Washington, D.C., 1978.
46. Hentrich, R. L., Grawes, G. A., Stein, H. G., and Bajpai, P. K., An evaluation of inert and resorbable ceramics for future clinical orthopedic applications, *J. Biomed. Mater. Res.*, 5, 25, 1971.
47. Hulbert, S. F., Young, F. A., Mathews, R. S., Klawitter, J. J., Talbert, C. D., and Stelling, F. H., Potential of ceramic material as permanently implantable skeletal prostheses, *J. Biomed. Mater. Res.*, 4, 433, 1970.
48. Hulbert, S. F., Morrison, S. J., and Klawitter, J. J., Tissue reactions to three ceramics of porous and non-porous structures, *J. Biomed. Mater. Res.*, 6, 347, 1972.
49. Hulbert, S. F., Cooke, F. W., Klawitter, J. J., Leonard, R. B., Sauer, B. W., Moyle, D. D., and Skinner, H. B., Attachment of prostheses to the musculoskeletal system by tissue ingrowth and mechanical interlocking, *J. Biomed. Mater. Res. Symp.*, 7, 1, 1973.
50. Jahnke, K. and Plester, D., Keramik-Implantate in der Mittelohrchirurgie, *HNO*, 28, 109, 1980.
51. Jahnke, K. and Plester, D., Praktische Hinweise zur Anwendung von Mittelohr-Implantaten aus Aluminiumoxid-Keramik, *HNO*, 28, 115, 1980.
52. Jarcho, M., Kay, S. F., Gumaer, K. J., Doremus, R. H., and Drobeck, H. P., Tissue, cellular and subcellular events at a bone-ceramic hydroxylapatite interface, *J. Bioeng.*, 1, 79, 1977.
53. Jarcho, M., Jasty, V., Gumaer, K. I., Katy, J. F., and Doremus, R. H., Electron microscopic study of a bone-hydroxyapatite implant interface, *Trans. 4th Ann. Met. Soc. Biomater. 10th Int. Biomater. Symp.*, 1978, 112.
54. Jarcho, M., Calcium phosphate ceramics as hard tissue prosthetics, *Clin. Orthop.*, 157, 259, 1981.
55. Jarcho, M., Bolen, C. H., Thomas, M. B., Bobick, J., Kay, J. F., and Doremus, R. H., Hydroxyapatite synthesis and characterization in dense polycrystalline form, *J. Mater. Sci.*, 11, 2027, 1976.

56. Jarcho, M., Salsbury, R. L., Thomas, M. B., and Doremus, R. H., Synthesis and fabrication of β-tricalcium phosphate ceramics for potential prosthetic applications, *J. Mater. Sci.*, 14, 142, 1979.
57. Jasty, V., Jarcho, M., Gumaer, K. I., Sauerschell, R., and Drobeck, H. P., Bone tissue response to dense hydroxyapatite disc implants in mongrel dogs. A light and electron microscopic study, *9th Int. Congr. Electron. Microsc.*, 2, 674, 1978.
58. Karbe, E., Köster, K., Kramer, H., Heide, H., Kling, G., and König, R., Knochenwachstum in porösen, keramischen Implantaten beim Hund, *Langenbecks Arch. Chir.*, 338, 109, 1975.
59. Kato, K., Aoki, H., Tabata, T., and Ogiso, M., Biocompatibility of apatite ceramics in mandibles, *Biomater. Med. Devices Artif. Organs*, 7, 291, 1979.
60. Katz, J. L., Basic engineering language for the mechanical properties of materials, paper presented at the 10th Int. Biomater. Symp., San Antonio, Tex., 1978, 16.
61. Kawahara, H., Yamagami, A., Nahamura, M., Imai, K., and Ohta, Y., Biological evaluation of implant materials cell adhesion of material, paper presented at the 10th Int. Biomater. Symp., San Antonio, Tex., 1978, 15.
62. Kay, J. F., Doremus, R. H., and Jarcho, M., Ion micromilling of bone-implant interfaces, *Trans. 4th Ann. Meet. Soc. Biomater., 10th Int. Biomater. Symp.*, 1978, 154.
63. Klawitter, J. J. and Hulbert, S. F., Application of porous ceramics for the attachment of load bearing internal orthopedic applications, *J. Biomed. Mater. Res. Symp.*, 2, 161, 1971.
64. Klein, C. P. A. T., van Kamp, G., de Groot, K., and Vermeiden, J. P. W., Possible involvement of the immune system in the degradation of hydroxylapatite implants, in *Dental Implants*, Heimke, G., Ed., Carl Hanser Verlag, Munich, 1980, 41.
65. Köster, K., Karbe, F., Kramer, H., Heide, H., and König, R., Experimental bone replacement with resorbable calcium phosphate ceramic, *Langenbecks Arch. Chir.*, 341, 77, 1976.
66. Köster, K., Heide, H., and König, R., Histologische Untersuchungen an der Grenzfläche zwischen Knochengewebe und Calziumphosphat-, Calziumaluminat- und Aluminiumoxidkeramik, *Z. Orthop.*, 115, 693, 1977.
67. Köster, K., Heide, H., and König, R., Resorbable calcium phosphate ceramics under load, *Langenbecks Arch. Chir.*, 343, 173, 1977.
68. Krempien, B., Griss, P., Heimke, G., von Andrian-Werburg, H., Hartung, H. J., Reipa, S., Lauterbach, H. J., Remodelling dynamics of cortical bone under the influence of ceramic implants in sheep, *Acta Orthop. Belg.*, 40, 624, 1974.
69. Krempien, B., Schulte, W., Kleineikenscheidt, H., Lindner, K., Shareyke, R., and Heimke, G., Lichtoptische und rasterelektronenmikroskopische Untersuchungen den der Grenzfläche von Implantaten aus Aluminiumoxid-Keramik im Unterkieferknochen von Hunden, *Dtsch. Zahnäerztl. Z.*, 33, 332, 1978.
70. Mooney, R. W. and Aia, M. A., Alkaline earth phosphates, *Chem. Rev.*, 61, 433, 1961.
71. Nery, E. B., Lynch, K. L., and Rooney, G. F., Alveolar ridge augmentation with tricalcium phosphate ceramic, *J. Prosthet. Dent.*, 40, 668, 1978.
72. Lemons, J. E., Ballard, J. B., Culpepper, M. I., and Niemann, K. M. W., Porous tricalcium phosphate ceramic for segmental lesions, paper presented at the 1st World Biomat. Congr., Baden, Austria, 4.10.3 (Abstr.), 1980.
73. Levin, M. P., Getter, L., Adrian, J., and Cutright, D. E., Healing of periodontal defects with ceramic implants, *J. Clin. Periodont.*, 1, 197, 1974.
74. Levin, M. P., Getter, L., and Cutright, D. E., A comparison of illiac marrow and biodegradable ceramic in periodental defects, *J. Biomed. Mater. Res.*, 9, 183, 1975.
75. McDavid, P. T., Boone, M. E., Kafrawy, A. H., and Mitchell, D. F., Effect of autogenous marrow and calcitonin on reactions to a ceramic, *J. Dent. Res.*, 58, 1478, 1979.
76. Mors, W. A. and Kaminski, E. J., Osteogenic replacement of tricalcium phosphate ceramic implants in the dog palate, *Arch. Oral Biol.*, 20, 365, 1975.
77. Nery, E. B., Lynch, K. L., Hirthe, W. M., and Mueller, K. H., Bioceramic implants in surgically produced infrabony defects, *J. Periodontol.*, 46, 328, 1975.
78. Nery, E. B., Pflughoeft, F. A., Lynch, K. L., and Rooney, G. E., Functional loading of bioceramic augmented alveolar ridge, *J. Prosthet. Dent.*, 43, 338, 1980.
79. Niwa, S., Sawai, K., Takahashi, S., Tagai, H., Ono, M., and Fukuda, Y., Experimental studies of the implantation of hydroxyapatite in the medullary canal of rabbits, paper presented at the 1st World Biomat. Congr., Baden, Austria, 4.10.4 (Abstr.), 1980.
80. Nothdurft, H., Tumorerzeugung durch Fremdkörperimplantation, *Abh. Dtsch. Akad. Wiss. Berlin Kl. Med.*, 80, 1960.
81. Ogiso, M., Kaneda, H., Arasaki, J., Tabata, T., and Hidaka, T., Epithelial attachment and bone tissue formation on the surface of hydroxyapatite ceramics, paper presented at the 1st World Biomater. Congr., Baden, Austria, 4.1.5 (Abstr.), 1980.
82. Oppenheimer, B. S., Oppenheimer, E. T., Danishefsky, J., Stout, A. P., and Eirich, F. R., Further studies of polymers as carcinogenic agents in animals, *Cancer Res.*, 15, 333, 1955.

83. Ott, G., Frendkörpersarkome, in *Experimentelle Medizin Pathologie und Klinik*, Vol. 32, Springer-Verlag, Basel, 1970.
84. **Peelen, J. G. J., Rejda, B. V., Vermeiden, J. P. W., and de Groot, K.**, Sintered tricalcium phosphate as bioceramic, *Sci. Ceram.*, 9, 226, 1977.
85. **Urist, M. R. and Strates, B. S.**, Bone morphogenetic protein, *J. Dent. Res.*, 50 (Suppl.), 1392, 1971.
86. **Plenk, H.**, Evaluation of the effect of ceramic and different metallic implants on the growth rate of human fibroblast cultures, *Evaluation of Biomaterials*, Winter, G. D., Leray, J. L., and de Groot, K., Eds., John Wiley & Sons, New York, 1980, 399.
87. **Polack, F. M. and Heimke, G.**, Ceramic keratoprostheses, *Opthalmology*, 87, 693, 1980.
88. **Predecki, P., Stephan, J. E., Auslaender, B. A., Mooney, V. L., and Kirkland, K.**, Kinetics of bone growth into cylindrical channels in aluminium oxide and titanium, *J. Biomed. Mater. Res.*, 6, 375, 1972a.
89. **Predecki, P., Auslaender, B. A., Stephan, J. E., Mooney, V. L., and Stanitzki, C.**, Attachment of bone to threaded implants by ingrowth and mechanical interlocking, *J. Biomed. Mater. Res.*, 6, 401, 1972b.
90. **Reyda, B. V.**, Composite Materials for Hard Tissue Replacement, Thesis, Free University, Amsterdam, 1977.
91. **Sayler, K., Holmes, R., and Johns, D.**, Replamineform porous hydroxyapatite as bone substitute in craniofacial osseous reconstruction, *J. Dent. Res.*, 56B, 173, 1977.
92. **Schulte, W.**, Die Reaktion des Blutgerinnsels und ihre Bedeutung für die primäre Heilung von Kieferknochendefekten, Thesis, University of Tübingen, 1962.
93. **Schulte, W., Kleineikenscheidt, H., Lindner, K., Shareyka, R., Heimke, G., Gerlach, Ch., and Hardegg, W.**, Tierexperimente zur Frage der Einheilung des Tübinger Sofortimplantates, *Dtsch. Zahnaerztl. Z.*, 33, 326, 1978.
94. **Schulte, W., Büsing, C. M., d'Hoedt, B., and Heimke, G.**, Enosseous implants of aluminium oxide ceramics: a 5 years' study in humans, *Proc. Int. Conf. Implantol. Biomater. Stomatol.*, Japan Scientific Society, Tokyo, 1980, 157.
95. **Shima, T., Keller, J. T., Alvira, M. M., Mayfield, F. H., and Dunsker, S. B.**, Anterior cervical disectomy and interbody fusion. An experimental study using a synthetic tricalcium phosphate, *J. Neurosurg.*, 51, 533, 1979.
96. **Strunz, V., Männer, K., Gross, U. M., Hirsch, A., Deutscher, K., and Brömer, H.**, The mechanical strength of the bond between bone and implants of glass-ceramic with apatite, in Dental Implants Heimke, G., Ed., Carl Hanser Verlag, Munich, 1980, 27.
97. **Winter, M., Griss, P., de Groot, K., Tagai, H., Heimke, G., van Dijk, H. J. A., and Sawai, K.**, Comparative histocompatibility testing of seven calcium phosphate ceramics, *Biomaterials*, 2, 159, 1981.
98. **Wolf, H.**, Zur komplexen Wechselwirkung von Blut mit nicht biologischen Oberflächen, *Dtsch. Gesundheitswes.*, 36, 626(1), 717(2), 1981

Chapter 5

CERAMICS OF CALCIUM PHOSPHATES: PREPARATION AND PROPERTIES

Klaas de Groot

TABLE OF CONTENTS

I. INTRODUCTION

As shown in Chapter 1, many calcium phosphate salts exist that are present in or related to bone and tooth mineral. Stabilities at high temperatures vary considerably, depending on such variables as Ca/P ratio, water content, impurities, and temperature. At physiological temperatures and surroundings, the components in equilibrium may be quite different. Calcium phosphate salts are mostly available in the form of powders or small single crystals. Nature provides larger samples of some compositions. For implant purposes, however, processes yielding larger blocks are necessary. In this chapter various techniques as they are in use today will be discussed, as well as the mechanical and biological properties of the products.

The earliest applications of calcium phosphate salts were in the form of powders. Albee[1] used "triple calcium phosphate" ($Ca_2(PO_4)_2$), as a stimulus to osteogenesis, in rabbit bone and found a positive influence on healing of fractures. Albee also mentioned clinical use of this powder on human subjects. However, lack of mechanical integrity renders a powder, or a mix of powders with serum, rather useless in clinical practice. Although the surgeon is most interested in implants that can be shaped and hardened *in situ,* such preparations are not yet available for bioceramics. At the moment, preparation techniques are too complicated and time consuming for "bed-side" application. Hence, implants must be preformed elsewhere.

The following sections will review techniques for making blocks of calcium phosphate ceramic, large enough to be used clinically. With each technique the relevant biochemical data of the product shall be discussed.

II. BLOCKS OBTAINED BY SINTERING TECHNIQUES

A. Firing Wet Slurries

Ceramics are most often prepared by preforming a slurry of powder and liquid (mostly water) into a shape or a mold and increasing the temperature to a point where the individual powder particles fuse more or less together, depending on temperature and time.[2] The further the sinter process progresses, the less the void volume between the particles. The void volume, or porosity in the resulting block, can be expressed in a percentage, usually referred to as microporosity if the particle size is in the order of microns. A small percentage of microporosity corresponds with an advanced sintering state. Several methods exist to introduce larger pores of at least 100 μm that may allow tissue ingrowth:[3]

1. When the slurry is made of powder and an aqueous solution of hydrogen peroxide, heating at temperatures of 80°C and higher results in the production of oxygen bubbles due to dissociation of hydrogen peroxide. These bubbles escape the drying slurry and leave behind interconnecting pores, the size of which may exceed 100 μm, depending on bubble size, a factor which in turn is determined by rate of temperature increase and hydrogen peroxide content in the slurry. This method, developed by Ryshkewith,[4] has been modified by Rejda et al.[5,6] to obtain a porous, degradable calcium phosphate ceramic. Rejda used commercially obtained apatite powder that was fired at temperatures between 1100 to 1300°C. At 1300°C, Rejda found only a few percent of micropores (3%), and could introduce up to 70% of macropores (Figure 1).
2. A biological microstructure can be introduced when a skeletal microstructure is embedded in wax, treated with acid to remove the mineral, and the wax negative filled with slurry.[7] When the wax is burned off at 400°C, a porous positive in dried slurry is obtained. After firing, a positive replica in ceramic is obtained.

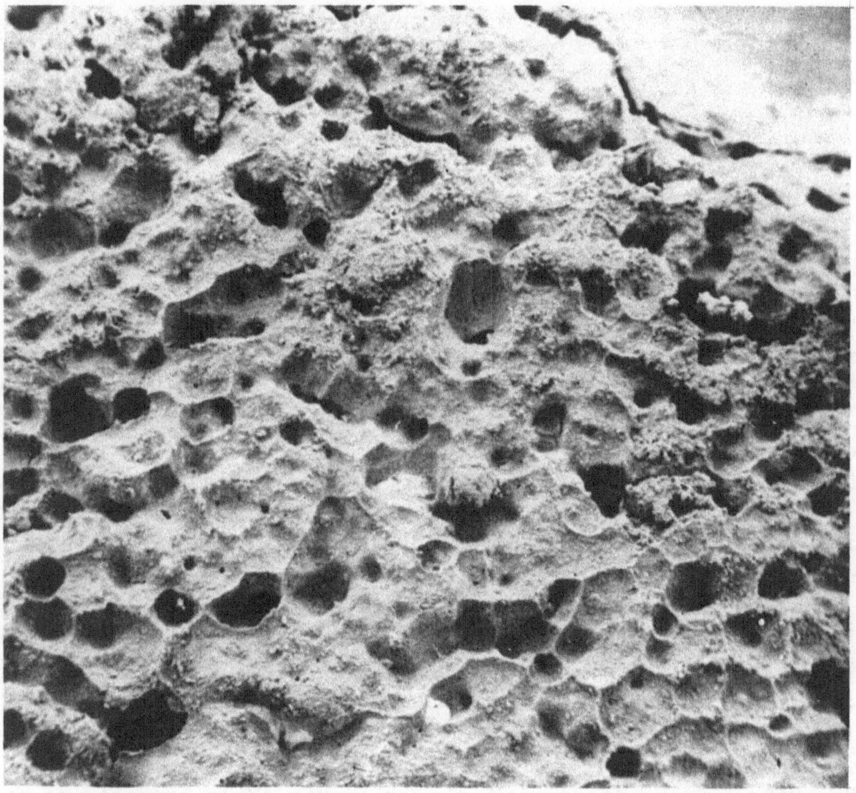

FIGURE 1. Surface of a macroporous sample of sintered apatite. The diameter of the pores is about 100 μm.

Instead of using a wax negative, any other porous structure that can be burned off at a lower temperature than the firing temperature can be used, such as naphthalene.[8] A very simple method, advocated by Köster et al.[9,10] is to drill holes mechanically into the green body (the state, obtained after the slurry has dried), the bisque (the state, when the sintering process has just begun), or the fired ceramic.

B. Firing Compacted Powders

Firing compacted powders has been the subject of many studies.[11-21] In this process, powders of calcium phosphate are compacted under pressure (up to 5000 kg/cm^2) in an ordinary press, or better yet, isostatically in a hydraulic press, and then sintered at the appropriate temperature. Several apatites and phosphates have been studied, such as fluorapatite, hydroxylapatite, tricalcium phosphate, dicalcium phosphate, and tetracalcium phosphate. The discussion shall be limited to calcium phosphates salts with Ca/P ratios between 1.50 and 1.67.

Very detailed studies on the relation of compacting techniques and resulting mechanical and structural properties such as microporosity have been published by Rootare.[17,22] It will be obvious that introduction of organic structures such as cellulose[21] or naphthalene[17] in the powder before compacting and sintering will result in blocks with macropores as desired. A method of theoretical interest is the so-called "hotpressing".[23] In this technique a powder is simultaneously compressed and fired, however, the high cost of the equipment prohibits laboratory use. For industrial applications the method might be attractive, especially when a modification, known as continuous hotpressing, is used.[24-26] An advantage is that dense materials with theoretically unlimited dimensions can be obtained.

III. STRUCTURAL AND MECHANICAL PROPERTIES

A. Introduction

In Chapter 1, the crystallographic properties of calcium phosphate salts prepared at high temperatures are discussed extensively. For the purpose of this chapter, the conclusions will be simplified as follows, realizing that exactness is not achieved by doing so.

In the range Ca/P ≈1.5 to Ca/P ≈1.7, two modifications, stable at high temperatures exist, hydroxylapatite and β-whitlockite. Apatite is more often expected when the Ca/P ratio is closer to 1.67 and a wet atmosphere is present, and β-whitlockite is found in a dry atmosphere and the Ca/P ratio is near 1.5. Starting powders, prepared by precipitation reactions at room temperature (at pH >4.2), always have an apatite structure. Ratios differing from 1.67 are accounted for by assuming defects, absorbed layers, and so on. When β-whitlockite is produced from such a powder, a phase transition must then occur from the original apatite structure. Obviously, this transition is promoted by Ca/P ratios near 1.5 and dry atmospheric conditions. It is known that an apatite starting powder with a Ca/P ratio of 1.5 changes to β-whitlockite at about 900 to 1000°C. On the other hand, as discussed in Chapter 1, the presence of Mg^{++} may destabilize the apatite lattice and favor a β-whitlockite structure. Hence, the conclusion is that sintered calcium phosphate salts, with a Ca/P ratio between 1.5 to 1.7 and sintered at a temperature higher than 1000°C, may have a bulk composition of β-whitlockite ("β-tricalcium phosphate") and/or apatite. When brought into contact with water, the surface will be changed to apatite, but when brought into contact with body fluids containing many additional components such as Mg^{++} salts, a whitlockite lattice can be found as well.

In any case, it seems plausible that calcium phosphate ceramics with Ca/P ratios between 1 and 2 are, at least at the surface, turned into physiological compositions when interactions with physiological fluids occur. As long as physiological compositions of bone are a matter of continuing discussion, we cannot expect otherwise from exact compositions of surfaces of calcium phosphate ceramics after implantation. Nevertheless, it may be concluded that after having reached an equilibrium state with body fluids, calcium phosphate ceramics will have a surface lattice, independent of an initial bulk Ca/P ratio. If, however, biological behavior is shown to depend on a Ca/P ratio, then the conclusion must be made that equilibrium states are not reached.

On the microscopical and macroscopical level, porosity plays an important role. It has already been seen that a sintered block may have macropores with a diameter of at least 100 μm to allow tissue ingrowth and micropores (space between individual particles). If a material has no pores at all, or less than a few volume percent (say 5%), it is called a dense material. Macropores or micropores may be present in blocks called macroporous or microporous, however, both types of pores can also be present. In that case, we can call the block porous. The classification of dense, microporous, macroporous and porous materials will be used throughout this monograph, as first delineated by Peelen et al.[27]

Mechanical strength depends on total porosity. In classical textbooks a linear relation between the log, σ, σ being strength, and V (volume fraction of both porosities) is suggested, the formula being:[28]

$$\log_e \sigma = \log_e \sigma (V = o) - n V$$

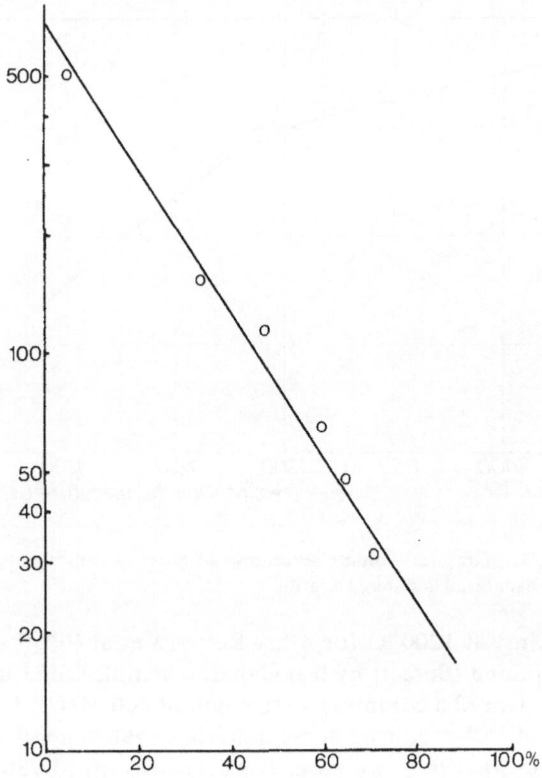

FIGURE 2. Graph of compressive strength (vertical axis, logarithmic scale in MN/m^2) vs. porosity in volume percentage.

For sintered apatites the data of Peelen et al.[27] and Denissen et al.[28,29] form a reasonably straight line, with a value of n = 3.8, in accordance with the value of n = 4−7 as given in the literature (Figure 2).[28] It is obvious that longer sinter times and higher sinter temperatures lead to reduction of microporosity, and hence to increase of strength. Figure 3 clearly shows this.[30]

B. Dense Materials

By drying and sintering of wet slurries, Jarcho et al.[31,36] succeeded in preparing dense apatite materials, with a density of almost 100%. Using a recipe of Hayek,[37] they dried a wet filtered cake of apatite powder and fired it at 1100 to 1200°C. Compressive strength is claimed to be up to 900 MN/m^2 and a fracture strength about 85 MN/m^2, but Akao et al.[38] showed the true compressive strength to be about 400 MN/m^2, in accordance with Figure 2. Jarcho[8] also developed a method of preparing dense β-whitlockite.

A Japanese group[38-40] also succeeded in a similar way as Jarcho to produce a dense apatite ceramic. Mechanical and crystallographic data were the same as found by Jarcho. Compacting and subsequent sintering of powders dates from an earlier period. Monroe et al.[41] used a commercial apatite powder with Ca/P = 1.54 and obtained a dense material that was chiefly dehydrated apatite and partly β-whitlockite. Monroe used cellulose to include macropores if wanted and measured the same hardness as natural dense apatite, but other properties are not given. Raja Rao and Boehm[42] sintered hydroxylapatite powder and obtained densities of about 95%, and compressive

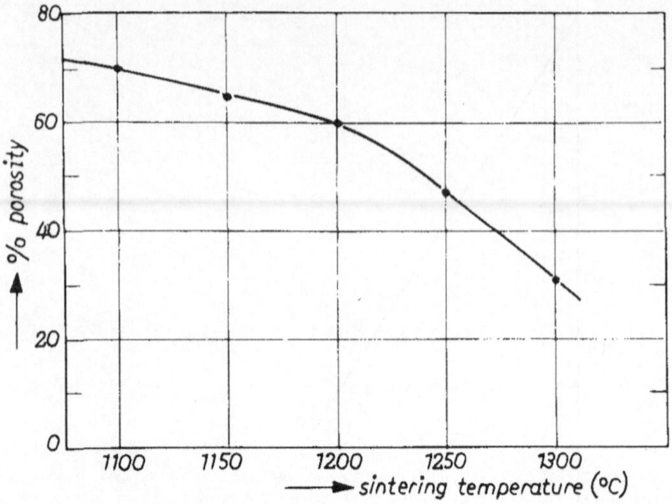

FIGURE 3. Graph of volume percentage of pores vs. sintering temperature. The material is sintered apatite.

strength of 200 MN/m^2 at 1200°C for 4 hr. Rootare et al.[17,22,43] used a powder with Ca/P = 1.5 and obtained (defect) hydroxylapatite samples, due to sintering in a wet atmosphere. They measured a compressive strength of 400 MN/m^2.

Later, Denissen et al.[29,44] produced dense materials (>90% apatite and <10% β-whitlockite), starting from apatite powders at temperatures up to 1300°C and 6 hr in a wet oxygen atmosphere. They found a compressive strength of 400 to 500 MN/m^2 and a tensile strength of 50 to 70 MN/m^2. Similar products were obtained by Denissen[26] when using the continuous hotpressing technique. Due to the low firing temperature (900°C) no β-whitlockite was formed. Biologically, there seems to be no difference between various types of densely sintered calcium phosphate blocks. Dense materials, tested in animals, are all apatitic with Ca/P ratios varying between 1.5 and 1.67. When implanted in bone, bone grows onto the implant surface, and no visible degradation takes place, as experiments in dogs, monkeys, and rats show.[31-36,45-46a] A tight bond bone implant is readily formed (Figure 4). For dental permucosal implantation it is interesting that reactions with gingival tissue resemble natural bonding tooth-tissue. According to Frank,[46b] hemidesmosomes are characteristic for the natural bond tooth-gingival tissue, and a Japanese group[46c] reported the same for the bond apatite implant gingival tissue. Similarly, bioglass with apatite coatings[46d] and permucosally loaded implants of apatite ceramics, showed a clinically "satisfying" bond at the permucosal site.[46e]

In conclusion, dense calcium phosphate materials have a compressive strength of about 600 MN/m^2, a tensile strength of about 50 to 80 MN/m^2, regardless of how they are produced. Depending on sintering conditions, apatite or β-whitlockite samples are obtained. That the end products do not only depend on a Ca/P ratio, but on sintering conditions as well, is shown by the fact that Rootare and associates, using a powder of Ca/P = 1.5, obtained only apatite, while Denissen, using a ratio of Ca/P = 1.67, reported a mixture of >90% apatite and <10% β-whitlockite. Also Monroe, starting with Ca/P = 1.54 chiefly produced (dehydrated) apatite. Generally, a β-whitlockite lattice seems to be favored at Ca/P = 1.5 and at dry sintering conditions, while apatite lattices are preferred at Ca/P ratios near 1.67 or at wet sintering conditions.

C. Porous Materials

Many articles have been published on materials that are not dense, but are either

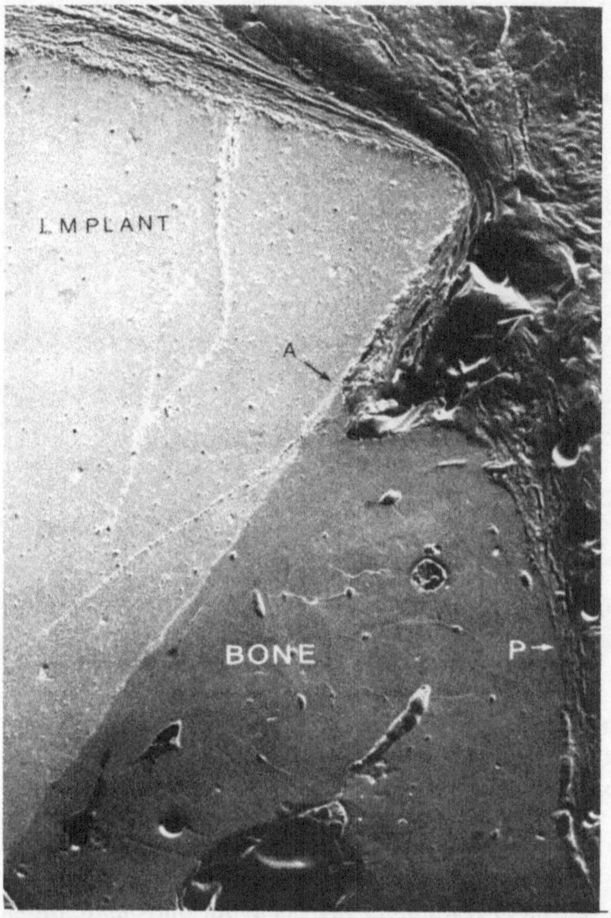

FIGURE 4. SEM photograph of interface alveolar ridge tissues (in dog) and implant surface (dense apatite).

microporous, macroporous, micro- and macroporous, or not specified with respect to porosity. As discussed in Section II.B., a variety of methods exists to produce blocks with porous structures. In the following section, the properties will be discussed with respect to preparation method.

1. Duplicating Biological Structures

So-called "replamineform implants" have been developed by applying either a method to duplicate the microstructure of acid-soluble components in ceramic[47,48] or a reaction to transform carbonate skeletal into apatites by a hydrothermal reaction:[49,50]

$$10\,CaCO_3 + 6(NH_4)_2HPO_4 + 2H_2O$$

$$Ca_{10}(PO_4)_6(OH)_2 + 6(NH_4)_2CO_3 + 4H_2CO_3$$

In accordance with Chapter 1, Mg-rich carbonates may turn into β-whitlockite structures. Mechanical strength is very low: 80 kg/cm² in compression.

Carbonate skeletals themselves resorbed readily after one year, and hydroxylapatite implants resorbed about 30% after the same time.[51-57] All studies are performed in dogs. Replamineform implants contain macropores as present in the duplicated tissue,

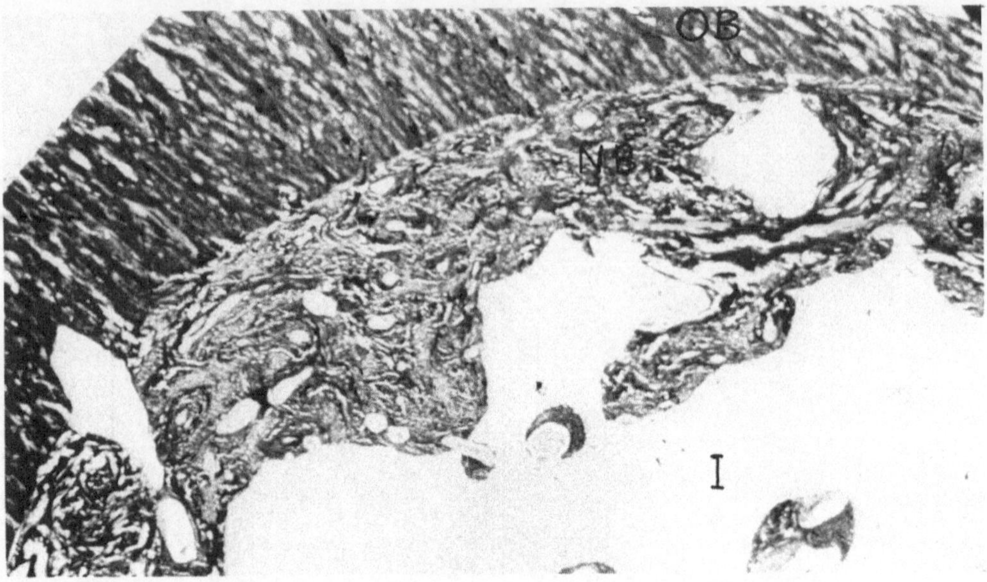

FIGURE 5. Resorbtion of porous implant of mixture of apatite (90%) and β-whitlockite. OB is rabbit bone in which round implant is inserted; NB is implant replaced by new bone; I is implant not yet degraded (after 3 months).

while no information is given with respect to microporosity. Without information on microstructure, it is not fully justified to conclude that calcium carbonate resorbs faster than hydroxylapatite.

2. Firing of Dried Slurries or Compacted Powders

Bhaskar[58] was the first to investigate porous blocks of calcium phosphate materials. Bhaskar reported on the degradability of tricalcium phosphate implants in rat tibiae. Numerous others confirmed this data, at least qualitatively,[59-70] on the material as supplied by Cameron et al.[71,72] who defined tricalcium phosphate as follows: strength is 20 to 70 kg/cm² (in compression, macropores of 100 to 300 μm are included), structurally it is β-whitlockite. No data on micropores are presented. Resorbtion is clearly cellular, although passive dissolution also undoubtedly takes place. Cells of mesenchymal origin,[58] macrophages,[65,66] and giant cells[71] are observed, sometimes with ceramic particles in their cytoplasm.[58,71] They replace the implanted block by bone. When implanted in rabbit ear, only resorbtion takes place without new bone formation.[73]

Various other groups experimented with calcium phosphate salts, which are not always clearly defined crystallographically or microstructurally. Biggs et al.[74,75] published materials on a calcium phosphate ceramic with Ca/P = 1.5 and found the implant to be resorbable in dogs and rabbits, the latter animal showing a more rapid resorption. A German group experimented with Ca/P ratios varying between 1 and 2, but it is not clear how microstructural properties vary within this range. They showed resorbtion up to Ca/P = 1.5. Strength varied from 1200 kg/cm² for their Ca/P = 1.5 sample to 1900 kg/cm² for a block with Ca/P = 2. Crystallographically, their tricalcium phosphate is β-whitlockite.[76-79] Rejda and Peelen et al.[5,6,27,80,81] developed a degradable material, based on commercially obtained tricalcium phosphate, with Ca/P = 1.67. Their end product is 90% apatite and 10% whitlockite, which contained controlled fractions of micro- and macropores. Compressive strengths of several thousand kg/cm² are obtained in accordance with Figure 2. They investigated blocks possessing both types of pores and found degradation. No studies delineating possible effects of each type of pores were conducted (Figure 5).

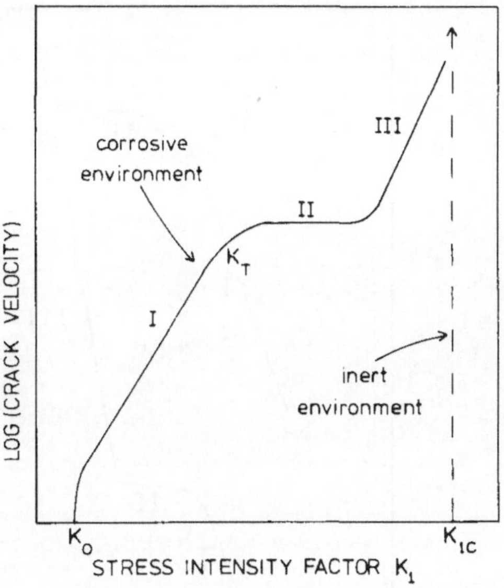

FIGURE 6. Typical relation between crack velocity and stress intensity factor. The material is glass.

D. Fatigue

1. Introduction

Failure by fatigue, an aspect of ceramics, has been ignored for some time in biomedical research until very recently.[93] For biomedical purposes, long-time duration strength is important. For example, if a glass rod breaks at a stress σ in a short-time stress, an identical rod when stressed to about 0.75 σ would fracture in a time about 100 times longer. Generally, decrease of (tensile) strength of brittle ceramic materials is caused by slow or subcritical crack growth, occurring under stress, sometimes assisted by environmental factors. Of great importance is the relationship between the so-called "stress-intensity factor" (K_I) and crack velocity ($\dot{a}$).

It is rather well-established that with increasing K_I, a different relation between a and K_I may exist. At very low K_I, sometimes called the threshold value K_{Io}, the relation may be simply $\dot{a} = o$ (or: $\dot{a} = K_{Io}^n$, $n = -\infty$), while at higher regions the approximate formulas are $\dot{a} = \alpha_1 K_1^n$ (n = 10 to 20 for oxides), $\dot{a} = \alpha_2$ (or $a = K_I^n$, n = o), and, at high K_I approaching K_{Ic}, the value where fracture is instantaneous, $\dot{a} = \alpha_3 K_I^n$ (n = 100, for all materials). These empirical formulas show that for ceramics, wide regions are adequately represented by simple power functions $\dot{a} = \alpha K_I^n$, where n is either in the order of 20, or approaching 100. Depending on which region dominates, subcritical crack growth, and hence fatigue failure, may be substantial or almost negligible. Figure 6 gives an example.[94] Material A, having a large region with n = 20, and a small region with n = 100, will have a more substantial crack growth $\dot{a}$, than material B, although K_{Ic}, where immediate fracture occurs, is the same for both materials. This means that, by equal K_{Ic}, a material with a large region with n $\approx$100, has much better fatigue resistance, than a material where the power relation is mainly $\approx$20. It is generally assumed that crack velocities with power relation n $\approx$20 are controlled by an interaction between ceramic and the water of the environment, while n $\approx$100 is associated with "intrinsic" behavior. Thus, a material in a dry surrounding is expected to have a larger region with n $\approx$100, than the same material in a humid environment. Figure 7 shows this clearly. It is obvious, that Figures 6 and 7 resemble each other, thus showing that

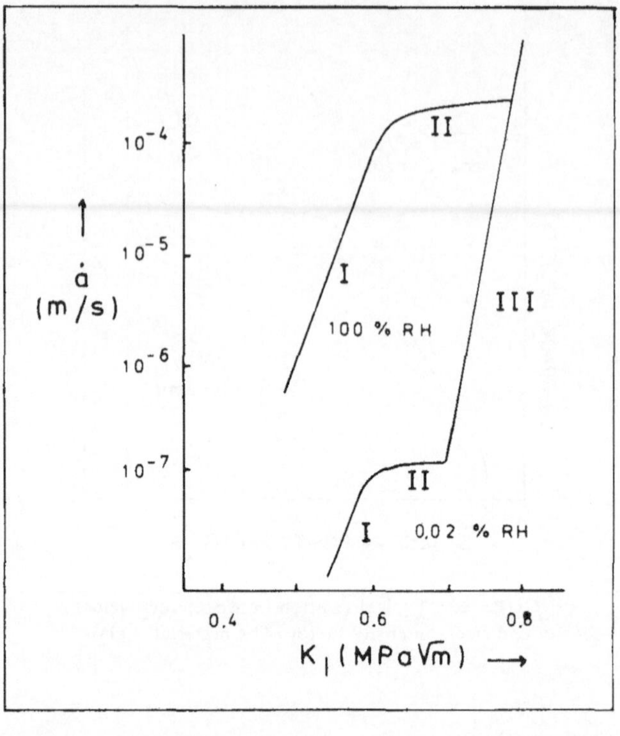

FIGURE 7. Same curves as in Figure 6. In 100% RH (relative humidity), much larger crack velocities are found than in 0.02% RH.

environment plays as important a role as material itself. Glass in a dry environment shows much less subcritical crack growth, and hence fatigue failure, than glass loaded in a humid atmosphere. This influence of humidity is very important with respect to bioceramics, being applied in physiological environments of 100% humidity and usually mechanically tested in dry environments. Mechanical data, not measured under physiological, or at least humid conditions, should be considered with care.

2. Fatigue of Calcium Phosphate Ceramics

Although aging and fatigue studies of porous calcium-aluminate ceramic implants have been studied in vivo a number of years ago (see Reference 1), indicating that this type of bioceramics can in essence be used only as a space filler, systematic studies on calcium phosphate implants have not been published. An exception is the report of Shima et al.,[97] who implanted porous β-whitlockite between vertebrae of rabbits. These researchers found essentially the same thing: that porous implants may fracture, and are then reduced to granulate.

Laboratory studies have been performed by de With et al.,[95] and by Thomas et al.[96] De With found that dry and dense apatite ceramics showed substantial subcritical crack growth, as evidenced by the value of 26 and 13, for dry (in dry nitrogen) and wet (in distilled water) samples, respectively. Thomas reported for dry samples an N ≈90, indicating less subcritical crack growth for dry samples. Since both groups report a similar ultimate tensile strength, it seems that preparation techniques do have an influence on fatigue resistance. In any case, however, the worrisome conclusion must be drawn that ceramics of even densely sintered calcium hydroxyapatite do have a limited resistance to fatigue failure. Since it is shown that in clinical practice even much stronger aluminumoxide ceramics may fail, the clinically relevant conclusion must be that apatite ceramics as such are not appropriate for permanent skeleton repair, if loading other than compression occurs.

IV. CONCLUSIONS

With respect to fabrication, conclusions are rather straightforward. As discussed in Section III.B., dense materials can be easily obtained by using commercial powders. A simple process of pressing and subsequent sintering suffices to produce a material with mechanical properties almost identical to those of dental enamel, a tissue which can be considered as dense apatite (97% theoretical density). Porous materials, with desired fractions of micropores and macropores, are as easily prepared, such that their mechanical properties are defined according to Figure 1. Compressive strength is equal to that of bone with comparable porosity, but the presence of organic matrix makes bone far superior in terms of tensile strength and impact strength. An example is that a calcium phosphate block with the same compressive strength as bone has only ½ its tensile strength, and ⅓ of its impact strength. Fatigue failure seems to limit the clinical possibilities to nonloaded or only compressed sites. With respect to crystalline structure, it can be concluded that either apatite or whitlockite materials can be prepared.

Biological properties are somewhat more difficult to assess in terms of microstructure (porosity), crystallinity (apatite vs. whitlockite), or other variables such as type of animal or place of implantation. These shall be discussed along with dense materials. All reports agree on the nonbiodegradability of dense apatite materials, as well as the strong bond between bone and implant. Dense whitlockite materials have not been biologically tested as yet. Hardly any encapsulations have been found, indicating the dense materials to be very biocompatible.

As described in Chapter 1, it can be expected that after incubation in physiological fluids, surfaces of apatite or whitlockite blocks will have the same surface (if equilibrium is reached) corresponding with the composition of the surrounding fluids and not the same as the exact chemical and crystallographical composition of the implant itself. In view of this conclusion, it does not seem surprising that interaction-dense apatite-gingival tissue resembles that of tooth-gingival tissue. However, a few reports[72c,82,83] have appeared recently suggesting that reaction to dense apatites is not as homogenous as previously suggested. The same material,[72c,82] or differently prepared dense materials,[83] show the bone implant layer to vary in thickness, composition, and local resorbtion behavior. The question arises whether this is due to variations in properties of the material surface or to a local difference in tissue reaction. Although it is tempting to consider material properties, it is useful to consider variations in natural interface behavior. It is well-known, for example, that interactions between tooth and alveolar bone vary widely even within the same individual. In dogs, inflammations are as frequently observed at the interface alveolar bone-apatite implant as at a natural interface alveolar bone-tooth.[46e] However, the possibility that differences in sintering conditions that are not reflected in varying porosities or crystallographical structures may result in different biological properties cannot be excluded. Klein et al.[84] found that both sinter temperature and lattice structures elicited differences in protein absorbtion, showing the assumption of thermodynamical equilibrium at the implant surface to be incorrect.

With respect to porous materials the situation is less clear-cut. Agreement exists about the biocompatibility, as evidenced by bony ingrowth in the large pores (>100 μm), and the absence of encapsulation. If biodegradation occurs, it is generally assumed that besides passive dissolution cellular activities play a role. This in turn suggests that both macropores (for increasing surface area available for cellular interaction) and micropores (determining the ease of isolating individual particles, to be digested intracellularly) are necessary for a degradable calcium phosphate material. These porosities might be a necessity, but of course, not necessarily sufficient conditions for biodegradation.

Almost all calcium phosphate salts studied are found to degrade by some authors, and found to be nondegradable by others. Hubbard[85] found no degradation of $Ca_2P_2O_7$, while Köster et al.[9] did. Hubbard and Lemons[86] reported no degradation of tricalcium phosphate (presumably whitlockite), while many authors did. Jarcho[33] stated that (dense) apatite samples do not degrade, while Ray,[87] Rejda et al.,[5] and Holmes[88] found (porous) apatite samples degrade. Griss et al.[83] found both apatite and whitlockite samples degrade, but reported a lower degradation rate for apatite.

In view of the theoretical considerations of Chapter 1, surfaces of various calcium phosphate implants, when incubated in physiological fluids, will have a similar surface, only if equilibrium is reached. In this case chemical differences would not be expected as an explanation for variations in biological behavior. It seems then, at first sight, that differences in porosity are not only necessary, but also sufficient to explain differences in the biodegradation rate.[89] Recent data, however, seem to contradict this assumption.

When injecting apatite powders in rat synovial joints, Denko et al.[90] found an decrease in inflammation when powders had been heated to about 200°C, the usual sterilization temperature. In addition, it has already been mentioned that Klein et al.[84] investigated serum protein absorbtion to apatites, heated at various temperatures, and found indications that absorbtion of some protein depended strongly on sintering temperature. It will be clear, that the first step after inserting an implant will be the formation of a protein layer on the surface. Thus biocompatibility and biodegradation are properties determined by this layer. If these properties differ, the protein layers must differ, and therefore not only must the absorbtion behavior of calcium phosphate surfaces differ, but the nature of the calcium phosphate surfaces themselves as well.

It follows then, that the occurrence of differences in biological behavior between various calcium phosphate surfaces is not in agreement with the assumption that all calcium phosphate surfaces have the same surface after incubation in serum. Thus, protein absorbtion to the surface must proceed faster than the process of turning the surface into the thermodynamically stable phase. At first sight, one expects an equilibrium where small ions are involved changing calcium phosphate surfaces into bone mineral to be reached faster than the formation of an absorbtion layer where large, and therefore less mobile, protein molecules are involved. However, since this simple point of view cannot explain the various findings on biodegradation, we must assume that calcium phosphate implants with a surface structure different from the thermodynamically most favored situation do not reach, at least not always, an equilibrium with surrounding physiological fluids. Hence, we must conclude that although porosity plays an important role in biodegradation, crystallographic factors may be important as well with regard to the question whether degradation of the implant occurs.

Other well-known factors that influence bone remodeling in general, and may play a role, are especially the presence of trace elements such as fluor,[94] copper,[91] silicium,[92] and manganese, that are probably present in all calcium phosphate ceramics.

REFERENCES

1. **Albee, F. H.**, Studies in bonegrowth-triple calcium phosphate as a stimulus to osteogenesis, *Ann. Surg.*, 71, 32, 1920.
2. **Wyatt, O. H. and Dew-Hughes, D.**, *Metals, Ceramics and Polymers,* Cambridge University Press, London, 1974.
3. **Klawitter, J. J.**, A Basic Investigation of Bone Growth Into a Porous Material, Ph.D. thesis, Clemson University, Clemson, S.C., 1970.
4. **Ryshkewitch, E.**, Compression strength of porous sintered alumina and zirconia, *J. Am. Ceram. Soc.*, 36, 65, 1953.
5. **Rejda, B. V., Peelen, J. G. J., and de Groot, K.**, Tricalcium phosphate as a bone substitute, *J. Bioeng.*, 1, 93, 1977.
6. **Rejda, B. V., Peelen, J. G. J., Vermeiden, J. P. W., and de Groot, K.**, Botvervangend materiaal vervaardigd uit calciumfosfaat, *Ned. Tijdschr. Geneeskd.*, 122, 625, 1978.
7. **Dunn, E. J. and White, E. W.**, Bone ingrowth in replamineform porous implants, in *Biocompatibility of Implant Materials,* Williams, D., Ed., Sector Publishing, London, 1976.
8. **Jarcho, M., Salsburg, R. L., Thomas, M. B., and Doremus, R. H.**, Synthesis and fabrication of β-tricalcium phosphate (whitlockite) ceramics for potential prosthetic applications, *J. Mater. Sci.*, 14, 142, 1979.
9. **Köster, K., Karbe, E., Kramer, H., Heide, H., and König, R.**, Experimentelle Knochenersatz durch resorbierbare Calciumphosphat-Keramik, *Langenbecks Arch. Chir.*, 341, 77, 1976.
10. **Köster, K., Heide, H., and König, R.**, Histologische Untersuchungen an der Grenfläche zwischen Knochengewebe und Calciumphosphat-, Calciumaluminat-, und Aluminiumoxydkeramik, *Z. Orthop. Ihre Grenzgeb.*, 115, 693, 1977.
11. **Aoki, H., Kato, K., and Tabata, T.**, Osteocompatibility of apatite ceramics in mandibles, *Rep. Inst. Med. Dent. Eng.*, 11, 33, 1977.
12. **Biggs, A., Shepherd, N., Stanwich, L., and Doku, M. C.**, Inducing osseous proliferation with biodegradable ceramic implants, *J. Dent. Res.*, 53(Abstr.), 124, 1974.
13. **Cameron, H. U., MacNab, I., and Pilliar, R. M.**, Evaluation of a biodegradable ceramic, *J. Biomed. Mater. Res.*, 11, 179, 1977.
14. **Levitt, S. R., Crayton, P. H., Monroe, E. A., and Condrate, R. A.**, Forming methods for apatite prosthesis, *J. Biomed. Mater. Res.*, 3, 683, 1969.
15. **Clarke, W. J., Driskell, T. D., Hassler, C. R., Tennery, V. J., and McCoy, L. R.**, Calciumphosphate resorbable ceramics. A potential alternative to bone grafting, IADR 51st Gen. Meet., International Association of Dental Research, Washington, D.C., Abstr. 259, 1973.
16. **Raja Rao, W. and Boehm, R. F.**, A study of sintered apatites, *J. Dent. Res.*, 53, 1351, 1974.
17. **Rootare H. M. and Craig, R. G.**, Characterization of the compaction and sintering of hydroxyapatite powders by mercury porosimetry, *Powder Technol.*, 9, 199, 1974.
18. **Denissen, H. W., van Dijk, H. J. A., Gehring, A. P., and de Groot, K.**, Prevention of densely sintered calciumhydroxyapatite, IADR 57th Gen. Meet., Abstr. 613, International Association of Dental Research, Washington, D.C., 1979.
19. **Graves, G. A., Hentrich, R. L., Stein, H. G., and Bajpai, P. K.**, Resorbable ceramic implants, *J. Biomed. Mater. Res. Symp.*, 2, 91, 1971.
20. **Hubbard, W.**, Physiological calcium phosphate as orthopedic implant material, *Diss. Abstr. Int. B.*, 35, 1974.
21. **Monroe, E. A., Ward, V., Ward, D. B., and McMullen, J.**, New calcium phosphate ceramic material for bone and tooth implants, *J. Dent. Res.*, 50, 860, 1971.
22. **Rootare, H. M. and Craig, R. G.**, Characterization of HA powders and compacts at room temperature and after sintering at 1200°C, *J. Oral Rehab.*, 5, 293, 1978.
23. **Murray, P., Livey, D. T., and Williams, J.**, The hot pressing of ceramics, in *Ceramic Fabrication Processes,* Kingery, W. D., Ed., MIT Press, Cambridge, 147, 1958.
24. **Oudemans, G. J.**, A continuous hot-pressing technique, *Proc. Br. Ceram. Soc.*, 12, 83, 1969.
25. **Klein Wassink, R. J.**, Continuous hotpressing of reactive materials, *High Temp. High Pressures,* 3, 411, 1971.
26. **Denissen, H. W., van Dijk, H. J. A., de Groot, K., Klopper, P. J., Vermeiden, J. P. W., and Gehring, A. P.**, Biological and mechanical evaluation of dense calcium hydroxyapatite made by continuous hot pressing, in *Mechanical Properties of Biomaterials,* Hastings, G. W. and Williams, D. F., Eds., John Wiley & Sons, New York, 489, 1980.
27. **Peelen, J. G. J., Rejda, B. V., Vermeiden, J. P. W., and de Groot, K.**, Sintered tricalciumphosphate as bioceramics, *Sci. Ceram.*, 9, 226, 1977.

28. Kingery, W. D., *Introduction to ceramics*, John Wiley & Sons, New York, 1960.
29. **Denissen, H. W., van Dijk, H. J. A., Gehring, A. P., and de Groot, K.**, Preparation of densely sintered calcium hydroxyapatite, IADR 57th Gen. Meet. Abstr. 613, International Association of Dental Research, Washington, D.C., 1979.
30. **Peelen, J. G. J., Rejda, B. V., and de Groot, K.**, Preparation and properties of sintered hydroxylapatite, *Ceramurgia Int.*, 4, 71, 1978.
31. **Stirling Drug**, Ceramic material, English Patent 1.522.182, 1978.
32. **Jarcho, M., Bolen, C. H., Thomas, M. B., Bobick, J., Kay, J. F., and Doremus, R. H.**, Hydroxylapatite synthesis and characterization in dense polycrystalline form, *J. Mater. Sci.*, 11, 2027, 1976.
33. **Jarcho, M., Kay, J. F., Gumaer, K. I., Doremus, R. H., and Drobeck, H. P.**, Tissue, cellular and subcellular events at a boneceramic hydroxylapatite interface, *J. Bioeng.*, 1, 79, 1977.
34. **Jarcho, M., Kay, J. F., Gumaer, K. I., and Doremus, R. H.**, Tissue cellular and subcellular events at a bone-hydroxylapatite interface (Abstr.), 3rd Ann. Meet. Soc. Biomater., New Orleans, 1977.
35. **Jarcho, M., O'Connor, J. R., and Paris, D. A.**, Ceramic hydroxylapatite as a plaque growth and drug screening substrate, *J. Dent. Res.*, 56, 151, 1977.
36. **Jarcho, M., Jasty, V., Gumaer, K. I., Kay, J. F., and Doremus, R. H.**, Abstr. 4th Ann. Meet. Soc. Biomater., San Antonio, Tex., 1978.
37. **Hayek, E. and Newesely, H.**, Pentacalcium, monohydroxy orthophosphate, *Inorg. Synth.*, 7, 63, 1963.
38. **Akao, M., Iida, M., Ouchi, K., Aoki, H., and Kato, K.**, Mechanical properties of sintered hydroxylapatite (Abstr), 5th Ann. Meet. Soc. Biomater., Clemson, S.C., 1979.
39. **Aoki, H., Kato, K., and Tabata, T.**, Osteocompatibility of apatite ceramics in mandibles, *Rep. Inst. Med. Dent. Eng.*, 11, 33, 1979.
40. **Tagan, H. and Aoki, H.**, Preparation of synthetic hydroxyapatites and sintering of apatite ceramics, Abstr. Bioceramics Symp., Keele, England, 1978.
41. **Monroe, E. A., Ward, V., Bass, D. B., and McMullen, J.**, New calcium phosphate ceramic material for bone and tooth implants, *J. Dent. Res.*, 50, 860, 1971.
42. **Raja Rao, W. and Boehm, R. F.**, A Study of sintered apatites, *J. Dent. Res.*, 53, 1351, 1974.
43. **Rootare, H. M., Powers, J. M., and Craig, R. T.**, Sintered hydroxylapatite ceramics for wear studies, *J. Dent. Res.*, 57, 777, 1978.
44. **Denissen, H. W. and de Groot, K.**, The response of the apatite ceramic surface to a simulated physiological environment, in *Dental Implants*, Heimke, G., Ed., Carl Hanser Verlag, Munchen, 1980, 35.
45. **Boyne, P. J. and Shapton, B. A.**, The response of surgical periodontal defects to implantation with a HA ceramic (Abstr.), 4th Ann. Meet. Soc. Biomater., San Antonio, Tex., 1978.
46. **Denissen, H. W., de Groot, K., Makkes, P. Ch., van den Hooff, A., and Klopper, P. J.**, Tissue response to dense apatite implants in rats, *J. Biomed. Mater. Res.*, 14, 713, 1980.
46a. **Denissen, H. W., Veldhuis, A. A., Makkes, P. Ch., van den Hooff, A., and de Groot, K.**, Dense apatite implants in preventive prosthetic dentistry, *J. Clin. Prev. Dent.*, 2, 23, 1980.
46b. **Frank, R., Fiore-Donne, G., Cimasoni, G., and Matter, J.**, Ultrastructural study of epithelial and connective gingival attachment in man, *J. Periodontol.*, 45, 626, 1974.
46c. **Ogiso, M., Kameda, H., Arasaki, J., Tabata, T., and Hidaka, T.**, Epithelial attachment and bone tissue formation on the surface of hydroxylapatite ceramics for the purpose of dental implantation, 1st. World Congr. Biomater., Abstr. 4.1.5, 1980.
46d. **Ogiso, M., Ohuchi, F., and Hench, L. L.**, Compositional dependance of the formation of calcium phosphate films on bioglass, *J. Biomed. Mater. Res.*, 14, 55, 1980.
46e. **de Putter, C., de Groot, K., Sillevis Smitt, P. A. E., and van den Hooff, A.**, Dichtgesinterde calcium hydroxylapatiet implantation, transmucosaal toegepast, *Ned. Tijdschr. Tandheelkd.*, in press.
47. **White, R., Weber, J., and White, E.**, Replamineform: a new process for preparing porous ceramic, metal and polymer prosthetic materials, *Science*, 176, 922, 1972.
48. **Weber, J., White, E., and Lebiedzik, J.**, New porous biomaterials by replication of echinoderm skeletal microstructure, *Nature (London)*, 233, 337, 1971.
49. **Roy, D. M., Eysel, W., and Dinger, D.**, Hydrothermal synthesis of various carbonate containing calcium hydroxyapatites, *Mater. Res. Bull.*, 9, 35, 1974.
50. **Roy, D. M. and Linneham, S. K.**, Hydroxyapatite formed from coral skeletal carbonate by hydrothermal exchange, *Nature (London)*, 247, 220, 1974.
51. **White, E. W., Weber, J. N., Roy, D. M., Owen, E. L., Chiroff, R. T., and White, R. A.**, Replamineform porous biomaterials for hard tissue implant applications, *J. Biomed. Mater. Res. Symp.*, 6, 23, 1975.
52. **Chiroff, R. T., White, E. W., Weber, J. N., and Roy, D. M.**, Tissue ingrowth of replamineform implants, *J. Biomed. Mater. Res. Symp.*, 6, 29, 1975.

53. Roy, D. M., Bones from the laboratory, *New Sci.*, 163, 1976.
54. Mitchell, O. G., Singh, I. F., Klein, I., and Roy, D. M., Histological evaluation of a porous HA implant in rat bone, *J. Dent. Res.*, 55, B243, 1976.
55. Holmes, R. E., Salyer, K. E., Tarhay, L., and White, E. W., Reconstruction of the canine mandible with replamineform hydroxyapatite (Abstr.), 3rd Meet. Soc. Biomater., New Orleans, 1977.
56. White, R. A., White, E. W., and Nelson, R. J., Uniform microporous biomaterials prepared by the replamineform technique, *Biomater. Med. Devices Artif. Organs*, 7, 127, 1979.
57. Holmes, R. E., Bone regeneration within a coralline hydroxyapatite implant, Plast. Reconstr. Surg., 63, 626, 1979.
58. Bhaskar, S. N., Brady, J. M., Getter, L., Gromer, M. F., and Driskell, T., Biodegradable ceramic implants in bone, *Oral Surg.*, 32, 336, 1971.
59. Clarke, W. J., Driskell, T. D., Hassler, C. R., Tennery, V. J., and McCoy, L. R., Calciumphosphate resorbable ceramics, a potential alternative to bone grafting, IADR 51st Gen. Meet., Abstr. 259, International Association for Dental Research, Washington, D.C., 1973.
60. Driskell, T. D., Hassler, C. R., Tennery, V. J., McCoy, L. R., and Clark, W. J., Calciumphosphate resorbable ceramics: a potential alternative to bone grafting, IADR 51st Gen. Meet., Abstr. 259, International Association for Dental Research, Washington, D.C., 1973.
61. Driskell, T. D., Hassler, C. R., and McCoy, L. G., The significance of resorbable bioceramics in the repair of bone defects, Proc. 26th Meet. ACEMB, 1973.
62. Tennery, V. J. and Driskell, T. D., SEM studies of bone ceramic interfaces in calcium phosphate resorbable ceramics, *Ceram. Bull. Soc. Sym. Biomater.*, 430, 1973.
63. Mors, W. A., Kaminski, E. J., Rosenstein, S., and Perry, H. T., Resorbable ceramic implants in surgically created cleft palates in dogs, *J. Dent. Res.*, 53 (Abstr.), 129, 1974.
64. Mors, W. A. and Kaminski, E. J., Osteogenic replacement of tricalcium phosphate ceramic implants in the dog palate, *Arch. Oral Biol.*, 20, 365, 1975.
65. Levin, M. P., Getter, L., Cutright, D. E., and Bhaskar, S. N., Biodegradable ceramic in periodontal defects, *Oral Surg.*, 38, 344, 1974.
66. Levin, M. P., Getter, L., Cutright, D. E., A Comparison of iliac marrow and biodegradable ceramic in periodontal defects, *J. Biomed. Mater. Res.*, 9, 183, 1975.
67. Niwa, S., Sawai, K., Takahashi, S., Tagai, H., Ono, M., and Fukuda, Y., Experimental studies on the implantation of HA in the medullary canal of rabbits, 1st World Biomater. Congr. Abstr. 4.10.4, 1980.
68. Streckbein, R. G., Wöltge, E., and Kubicek, J., Tierexperimentelle Erprobung von CaP-Keramik zum Verschluss von Knochendefekten in der faziaten Stirnhohlenwand sowie in der Kalotte des Hirnschadels, *Arch. Otol. Rhinol. Laryngol.*, 228, 85, 1980.
69. Ferraro, J., App, G., and Foreman, D., Analysis of $Ca_3(PO_4)_2$ bone implant in ilium canis to assess resorption, IADR 57th Gen. Meet., Abstr. 1278, International Association for Dental Research, Washington, D.C., 1979.
70. Ferraro, J. W., Experimental evaluation of ceramic calcium phosphate as a substitute for bone grafts, *Plast. Reconstr. Surg.*, 63, 634, 1979.
71. Cameron, H. U., MacNab, I., and Pilliar, R. M., Evaluation of a biodegradable ceramic, *J. Biomed. Mater. Res.*, 11, 179, 1977.
72. Cameron, H. U., Evaluation of a biodegradable ceramic (Abstr.), 25th Annu. ORS, San Francisco, 250, 1979.
73. Howden, G. F., Tissue reaction to the bioceramic synthos, BES 71st Sci. Meet., 3rd Conf. Mater. Use Med. Biol., Keele, England, 1978.
74. Biggs, A., Shepherd, N., Stanwich, L., and Doku, M. C., Inducing osseous proliferation with biodegradable ceramic implants, *J. Dent. Res.*, 53(Abstr.), 124, 1974.
75. Shepherd, N., Maloney, P., Biggs, A., Stanwich, L., and Doku, H. C., Subcortical ceramic implants for evaluation of alveolar ridges, *J. Dent. Res.*, 558, 244, 1976.
76. Köster, K., Karbe, E., Kramer, H., Heide, H., and König, R., Experimentelle Knochenersatz durch Resorbierbare Calciumphosphat-keramik, *Langenbecks Arch. Chir.*, 341, 77, 1976.
77. Köster, K., Heide, H., and König, R., Histologische Untersuchungen an der Grenzfläche zwischen Knochengewebe und Calciumphosphat-Calciumaluminat und Aluminiumoxidkeramik, *Z. Orthop.*, 115, 693, 1977.
78. Köster, K., Heide, H., and König, R., Resorbierbare Calciumphosphatkeramik im Tierexperiment under Belastung, *Langenbecks Arch. Chir.*, 343, 173, 1977.
79. Karbe, E., Köster, K., Kramer, K., Heide, H., Kling, G., and König, R., Knochenwachstum in porösen keramische Implantaten beim Hund, *Langenbecks Arch. Chir.*, 338, 109, 1975.
80. Peelen, J. G. J., Rejda, B. V., and de Groot, K., Preparation and properties of sintered hydroxylapatite, *Ceramurgia Int.*, 4, 71, 1980.
81. Peelen, J. G. J., Rejda, B. V., and Vermeiden, J. P. W., Sintered hydroxylapatite as a bioceramic, *Philips Tech. Rev.*, 37, 234, 1977.

82. Ducheyne, P. and de Groot, K., in press.
83. Griss, P., Winter, M., Heimke, G., and Sawai, K., Comparative biocompatibility testing of four different apatite products for bone replacement, 1st World Biomater. Congr., Abstr. 4.10.5, 1980.
84. Klein, C. P. A. T., de Groot, K., Vermeiden, J. P. W., and van Kamp, G., Interaction of some serum proteins with hydroxylapatite and other materials, *J. Biomed. Mater. Res.*, 14, 705, 1980.
85. Hubbard, W., Physiological calcium phosphate as orthopedic implant material, *Diss. Abstr. Int. B.*, 35, 1974.
86. Lemons, J. E., Ballard, J. B., Culpepper, M. I., and Niemann, K. M. W., Porous tricalcium phosphate ceramic for segmental bone lesions, Abstr. 4.10.3, 1st World Biomater. Congr., Abstr. 4.10.3, 1980.
87. Ray, R. D., A preliminary report on studies of basic calciumphosphate in bone replacement, *Proc. 37th Clin. Congr. Am. Coll. Surg.*, W.B. Saunders, Philadelphia, 1952, 429.
88. Holmes, R. E., Bone regeneration within a coralline hydroxyapatite implant, *Plast. Reconstr. Surg.*, 63, 626, 1979.
89. de Groot, K., Bioceramics consisting of calcium phosphate salts, *Biomaterials*, 1, 47, 1980.
90. Denko, C. W. and Petricevic, M., Hydroxyapatite crystal-induced inflammation and prostaglandin E_1, *J. Rheumatol.*, 6, 117, 1979.
91. Wilson, T., Katz, J. M., and Gray, D. H., Inhibition of active bone resorption by copper, *Calcif. Tissue Int.*, 33, 35, 1981.
92. Carlisle, E. M., Silicon: a possible factor in bone calcification, *Science*, 167, 179, 1970.
93. Hench, L. L., Bioceramics, *Sci. Ceram.*, 9, 193, 1977.
94. Okazaki, M., Moriwaki, Y., Aoba, T., Doi, Y., and Takahashi, J., Dissolution rate behavior of fluoridated apatite pellets, *J. Dent. Res.*, 60, 1907, 1981.
95. de With, G., van Dijk, H. J. A., Hattu, N., and Prijs, K., Preparation, microstructure and mechanical properties of dense polycrystalline hydroxyapatite, *J. Mater. Sci.*, 16, 1592, 1981.
96. Thomas, M. B., Doremus, R. H., Jarcho, J., and Salsbury, R. L., Dense hydroxylapatite; fatigue and fracture strength after various treatments, *J. Mater. Sci.*, 15, 891, 1980.
97. Shima, T., Keller, J. T., Alvira, M. M., Mayfield, F. H., and Dunsker, S. B., Anterior discectomy and interbody fusion: an experimental study using a synthetic tricalcium phosphate, *J. Neurosurg.*, 51, 533, 1979.

Chapter 6

DENTAL IMPLANTS

Klaas de Groot

TABLE OF CONTENTS

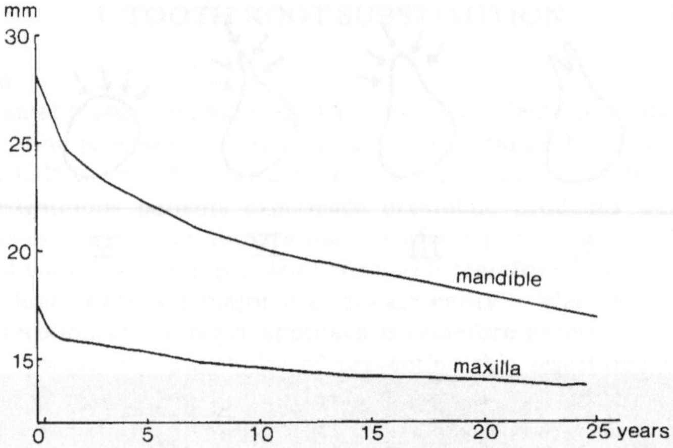

FIGURE 2. The graph shows loss of vertical height as a function of time of edentulousness.

An ingenious surgeon from America, lately settled in London, has satisfactorily shown that the deceased part of the tooth, including the cavity above described, may be cut off with the greatest ease and celerity, and that the sound root or roots may be allowed to remain in their sockets. This mode of procedure has been shown and explained to the most scientific surgeons, who have, without exception, expressed their conviction of its utility. The operation has been performed on more than two hundred members of the profession, and they have invariably declared that it occasioned no pain, and that it could be scarcely felt; as their own certificates show.]

What is of greatest moment, the stump or stumps left in the jaw afford a firm support to the adjoining teeth, and without which support, the alveolar process corresponding to the part before occupied by the diseased tooth and a part of the interstice structure of the jaw become absorbed; the adjoining teeth in a few years become loose, and ultimately prematurely fall out, as daily experience shows.

Mr. Fay has, for his improvement in this branch of surgery, and for other forceps for the perpendicular extraction of teeth, when extraction may be absolutely required, received the large silver medal from the Society of Arts.

FIGURE 3. The text of an article that appeared in the *Nottingham Review*, September 22, 1826. (From *Br. Dent. J.*, 4, 1980. With permission.)

undergoes very little resorbtion when well-fitting vitreous carbon was used as tooth root implants." Longer term results, however, show that most implants are rejected after longer times. Especially small dehiscencies are sufficient to exfoliate the implant,

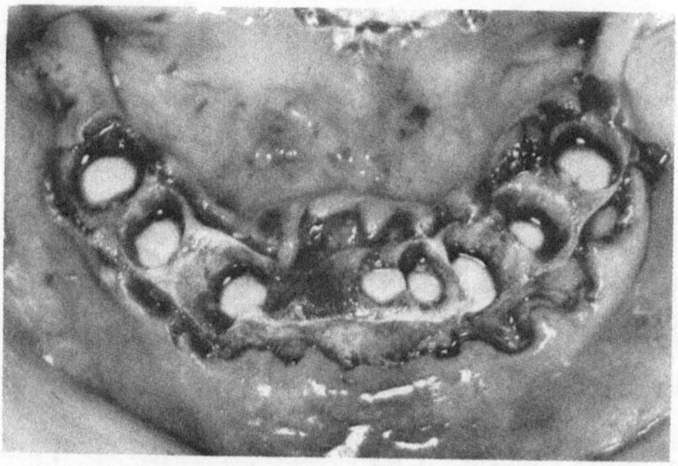

FIGURE 4. Eight apatite ceramic root implants after placement in the lower jaw of a patient.

while in general encapsulation occurs, frequently also gradually followed by exfoliation.[12,13]

Generally, the statement of Shulman and Schnitman[14] applies: natural tooth root implants are held by ankylosis, but fail due to resorbtion, while foreign materials are held by bone adaptation, but fail due to exfoliation.

C. Prevention of Residual Ridge Resorption By Implantation of Artificial Tooth Roots of Hydroxylapatite

As described by Denissen et al.,[15,16] implants of dense apatite, placed as immediate root substitute after extractions, succeed in forming a permanent part of the alveolar ridge. A follow-up of 2½ years showed that collapse of the ridge could be prevented when implants were placed under full lower dentures, fixed bridge work, and removable partial prosthetic constructions. Exposure of implants, occurring with a frequency of 3 to 4% per year under loaded full lower dentures, could be treated by renewed primary closure. Dense hydroxylapatite, prepared as described in Chapter 4, was shaped into truncated cones with lengths, upper, and lower diameters varying more or less according to root size. These roots were placed in the sockets, immediately after carefully extracting teeth that either were seriously periodontally involved or had draining fistula (Figure 4). Although the shape of the implants could be adapted more or less to the socket diameter, the fit of the implant was always very poor. It was sufficient to fix the implants in the sockets by lightly tapping with a mallet whereafter primary closure could follow. After a few months, radiographs showed the implants to be in close contact with the surrounding bone. No bone resorption was present in the periapical tissue (Figure 5). In Table 1 the results of 100 implants after 2½ years is presented. It is important to note that the implants were placed with their cervical plane situated just below the most apical part of the socket crest. This indicates that an implant may protrude from the alveolar contour after a few months, due to the rapid initial resorbtion occurring when the surrounding bone has not yet reached the implant. It is of primary interest to note that a close fit is not necessary, that dehiscencies can be treated easily, and that exposure does not lead to rejection. Dehiscencies leading to exposure, probably can be avoided when the implants are not placed at or just below the apical part of the socket crest, but 1 to 2 mm lower. In this case, the resorbtion during bone adaptation should lead to exposure of the implant less frequently.

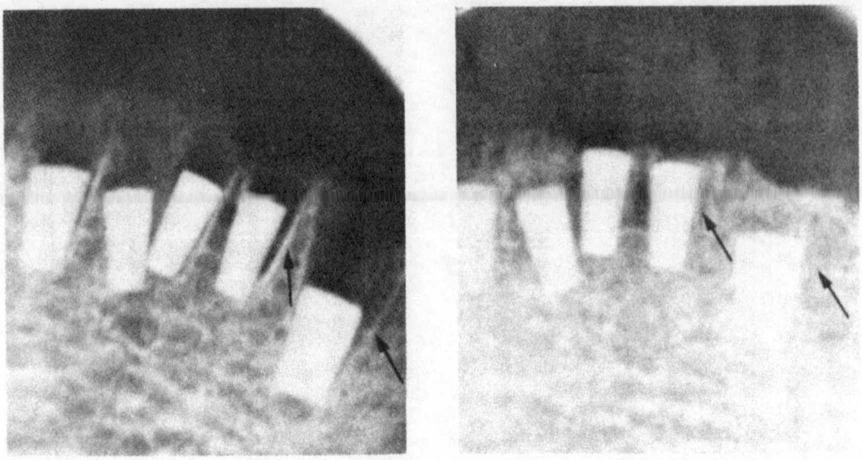

FIGURE 5. At left, radiograph immediately after implantation; at right, same patient after 25 months. Radiolucencies have disappeared. Lucency mesial of second implant of right is artefact caused by X-raying fused vestibular and lingual mucosal tissues.

Table 1
FATE OF TOOTH ROOTS OF SINTERED APATITE 2 YEARS AFTER IMPLANTATION IN MAN

	Implants under full lower dentures (71 implants in 9 patients)	All other implants (29 implants in 11 patients)
Absence of rejection phenomena	71	29
Retention	69	29
Exfoliation	2	0
Dehiscence	6	0
Preservation of the bulk of the ridge (clinical observation)	69	29

From Denissen, H. W., Veldhuis, A. A. H., Makkes, P. C. H., van der Hooff, A., and de Groot, K., *J. Clin. Prev. Dent.*, 2, 23, 1980. With permission.

D. Conclusion

Tooth root substitution by apatite implants seems to be a simple method of preventing the ridge to resorb after extraction of teeth (Figure 6). Implants do not need to fit accurately and rejection after occasional exposures under full lower dentures hardly occur. Dehiscencies can be treated easily. A slight change of implantation technique might reduce the dehiscencies and are estimated by Dennisen et al.[16] to be 3 to 4% per year. This means that a patient having 5 to 10 implants underneath a lower denture would have one dehiscency every three years. This seems to be acceptable, and hence a research program aimed at introducing this concept into routine dentistry should be pursued.

II. PERMUCOSAL IMPLANTS OF CALCIUM PHOSPHATE

A. Introduction

Permucosal implants are meant to transmit occlusal loads to the alveolar bone, and have therefore the primary task to avoid loads on the mucoperiosteum. Patients are

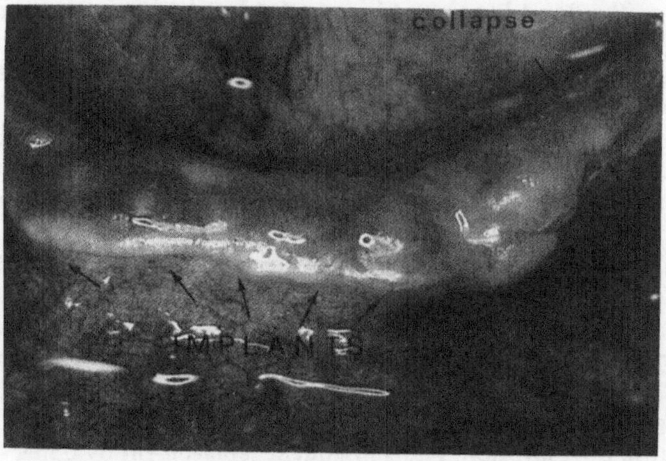

FIGURE 6. Implants prevent collapse of the cortical plates.

usually very interested in these types of implants, and the resulting commercial consequences have led to many premature devices. In a recent meeting of the Council on Dental Materials and Devices of the American Dental Association, a consensus was reached on the status of various implants.[17] In this meeting, the so-called Harvard consensus conference, the following definition of a successful implant was agreed upon: "The dental implant, to be considered successful, should provide functional service for five years in 75% of the cases." This is a rather modest requirement, if one considers Cranin's[18] finding, that various type of implants, regardless of shape and design, have a survival rate of 50% after 5 years. Thus, achieving such a survival rate is a lower limit, and not indicative for the quality of a design or material whatsoever.

In addition to a list of subjective criteria, the following objective criteria were included: "bone loss no greater than a third of the vertical height of the implant; good occlusional balance and vertical dimension; gingival inflammation amenable to treatment; mobility less than 1 mm in any direction; absence of symptoms and infection; absence of damage to adjacent teeth; absence of paresthesia or anesthesia or violation of the mandibular canal, maxillary sinus, or floor of the nasal passage; healthy collagenous tissue." Furthermore, data were considered statistically reliable if the number of implants is at least 40, and valid if representative of the general population. The conclusion reached at the conference was that, strictly speaking, none of the data on implants currently in use or advocated, were statistically valid. It was therefore agreed upon to warn against all data. Reports in the literature up to the time of this writing (mid-1981) have not essentially changed this situation. This also holds, of course, for permucosal implants of calcium phosphate, which will be discussed in the following paragraph.

B. Theoretical Considerations

Whether or not the objective criteria for a successful permucosal implant can be met depends on the answers to the following questions:[19] What is the nature of the bond between alveolar bone and implant when physiological loading, for example by chewing forces takes place? Does the permucosal site resemble the natural interface between mucosa and tooth? Can the material of which the implant is made withstand physiological loading?

1. The Nature of the Bond Between Alveolar Bone and Apatite Implant Under Physiological Loading

The nature of the bond between bone and apatite in general has been clearly described by various research groups. Since a detailed discussion is presented in Chapter 4, it is sufficient here to refer to the conclusion that under unloaded conditions an ankylosis is found. The question concerning how under oral conditions this ankylosis behaves, has been studied by de Putter et al.[19,20] These researchers implanted a number of apatite implants in predrilled holes in the jaw of beagles and studied the nature of the bond implant-bone (Figure 7). It was shown that no implant loosened after subjection to normal chewing forces (Figure 8). The dogs received normal dog food and were allowed to use their dentition freely. This animal study compares well with clinically known data on reimplanted teeth. As found by various authors, teeth reimplanted after treatments aimed at nonvitalizing the periodontal membrane are ankylosed to the surrounding alveolar bone. Although such teeth eventually fail after several years, they do so because of resorbtion, and subsequent replacement by bone, of the embedded root portion, but not to failure of the ankylosis under normally occurring occlusal and chewing forces.[14]

Therefore, it may be concluded that in human trials, occlusal and other physiological forces will not change the nature of the ankylosis found with apatite implants in alveolar bone under unloaded conditions.

2. The Nature of the Permucosal Site

For the permucosal site we can reach similar conclusions as to the bond between apatite implant material and surrounding bone: sufficient evidence exists to estimate the nature of the area when the apatite implants are in contact with soft oral tissues in humans. Again, clinical experiences on reimplantation of teeth may be discussed initially. The literature is unanimous in stating that attachment of gingival tissues to reimplanted teeth strongly resembles the natural gingival adherence. As shown by Frank[21] and other electron microscopical investigators,[22] attachment of gingival tissue to dentin and enamel is characterized by the presence of hemidesmosomes at the interface of soft tissue cells with tooth tissue.

Apatite implant materials show a similar behavior in dogs. A Japanese research team[23] has studied apatite ceramics of low strength to facilitate electron microscopical studies, and found mucosa adherence as evidenced by the presence of hemidesmosomes. The clinical aspects of mucosa at loaded implants was investigated by de Putter et al.,[19,20] and a modified Löwe index showed a behavior resembling the natural situation. These studies, on beagle dogs, showed dense forms of apatite to form an interface without unusual gingival inflammation and without symptoms and infections (Figure 9). The collagenous tissue seemed healthy. With respect to the permucosal site, we may conclude that sufficient evidence exists to reasonably assume that in humans apatite ceramics may fulfill the objective criteria of the Harvard Conference.

3. The Ability of Apatite Ceramics to Withstand Physiological Loading As Encountered in the Oral Cavity

A weak point of ceramics in general is their relative low tensile strength, and their low fatigue resistance to repetitive tensile loading.[24] Theoretical investigations show two important factors. First, that lifetimes in the real situation can hardly be predicted from laboratory experiments,[25] and second, that not only material properties, but also the environment is of utmost importance. For aluminum oxide ceramics, data show that in physiological solutions, lifetimes may be a factor of 100 lower than that in dry air.[26] The best test for a lifetime study is therefore the actual situation in humans, but a necessary prerequisite to do these are animal studies.

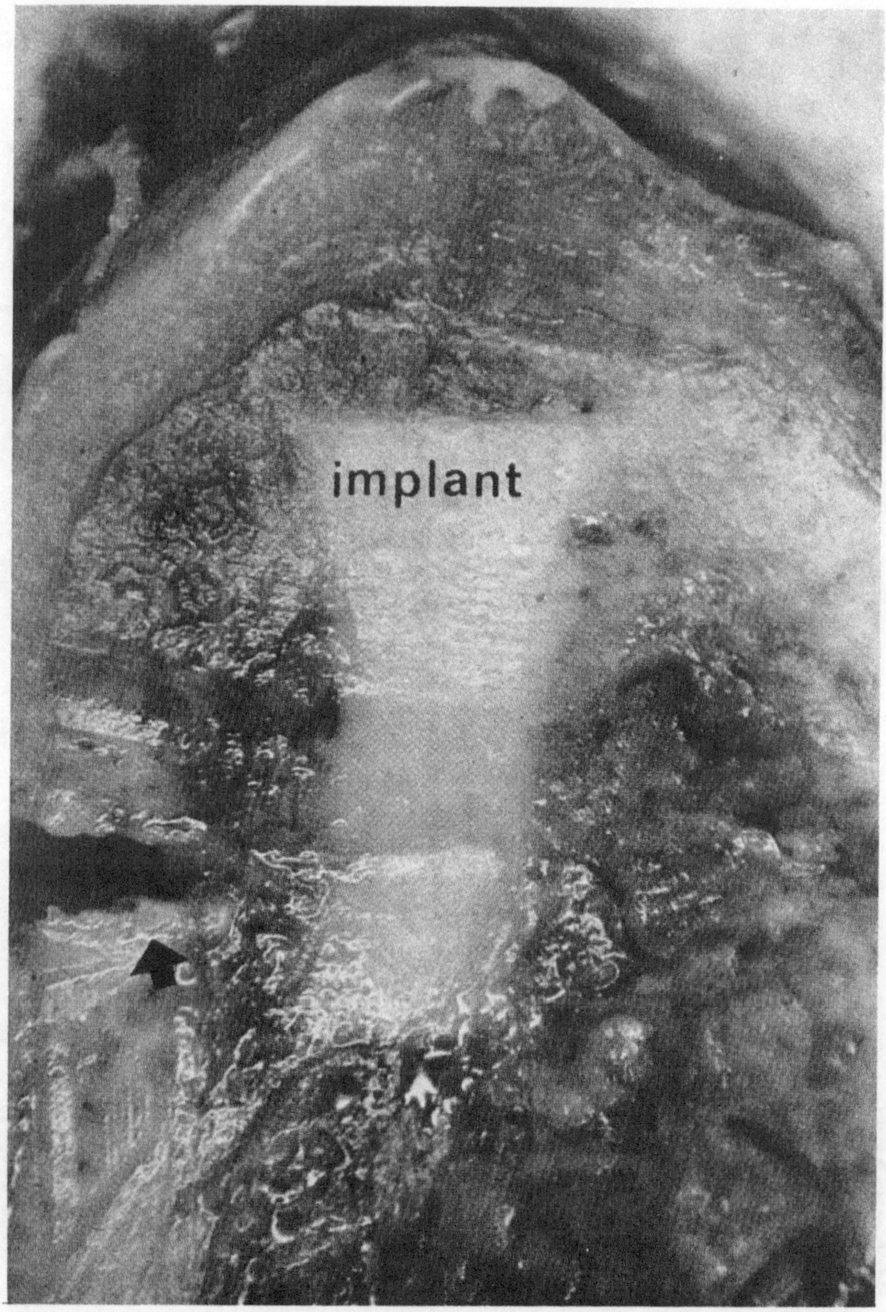

FIGURE 7. Surrounding bone is closely adapted to implant of sintered apatite in dog.

De Putter et al.[19,20] have implanted apatite ceramics in beagle dogs, and have shown that a number of implants do fail after a few months of functional use. They do not fail immediately, but rather suddenly after a given time after implants are fixed. Figure 10 shows limited fatigue resistance to be the main factor. Also, their data show that the shape of the implant is important. Different shapes have different lifetimes and survival rates.

Although animal studies yield neither a lower limit nor an upper limit for human

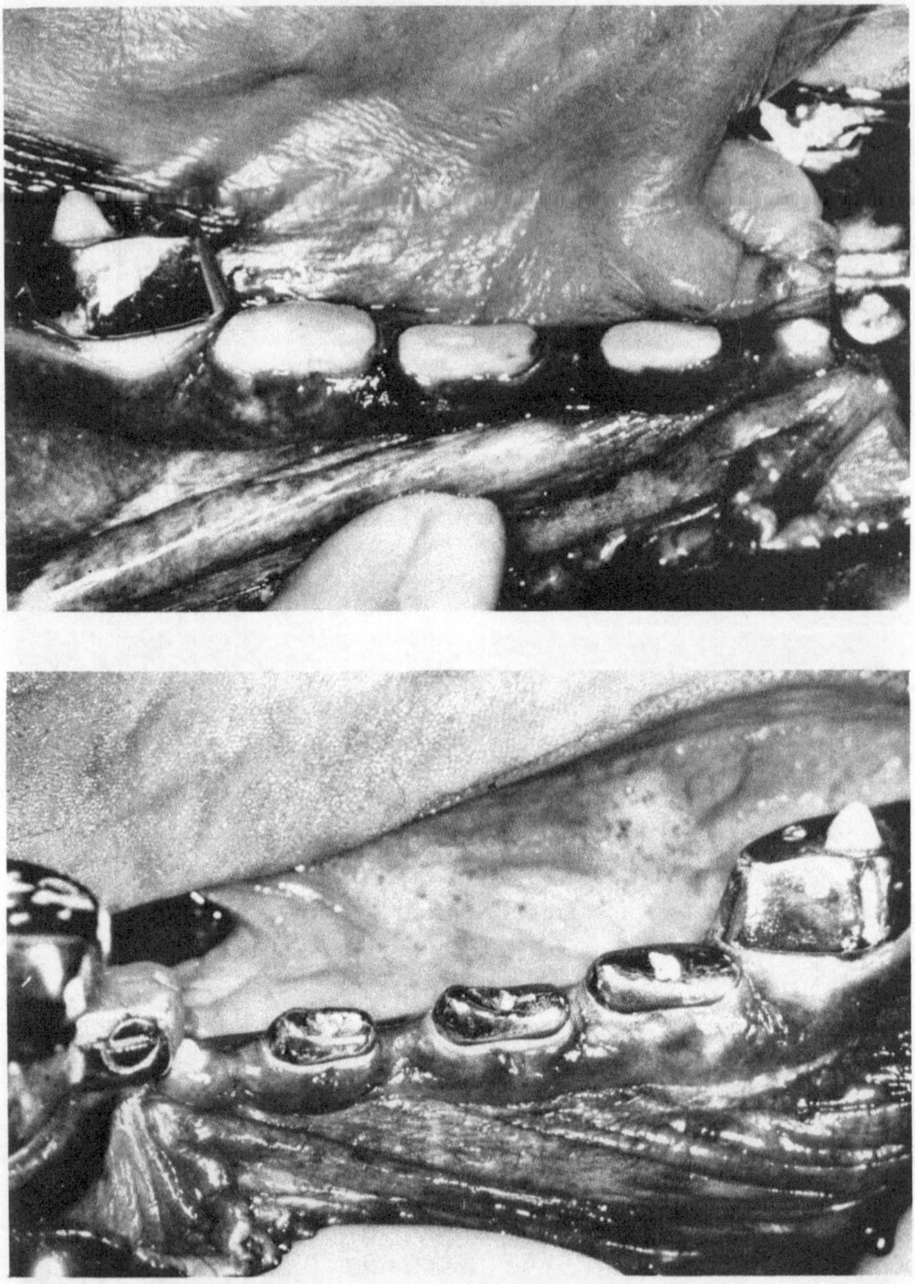

FIGURE 8. Top shows transmucosal implants of sintered apatite without crowns; bottom shows same implants with crowns (in dog, lower jaw).

applications, due to different loading conditions, it seems reasonable to conclude that fatigue properties of apatite ceramics are insufficient for use as a loaded transmucosal implant under all conditions. As discussed in Chapter 4, laboratory studies have revealed conclusive evidence for this statement. Therefore, only implants with limited or no occlusal contact and no tensile loading are indicated for human study. This leads inavoidably to a narrow indication field. A broader application can be expected when fatigue resistance could be improved. One then thinks of reinforcement by metals, thus

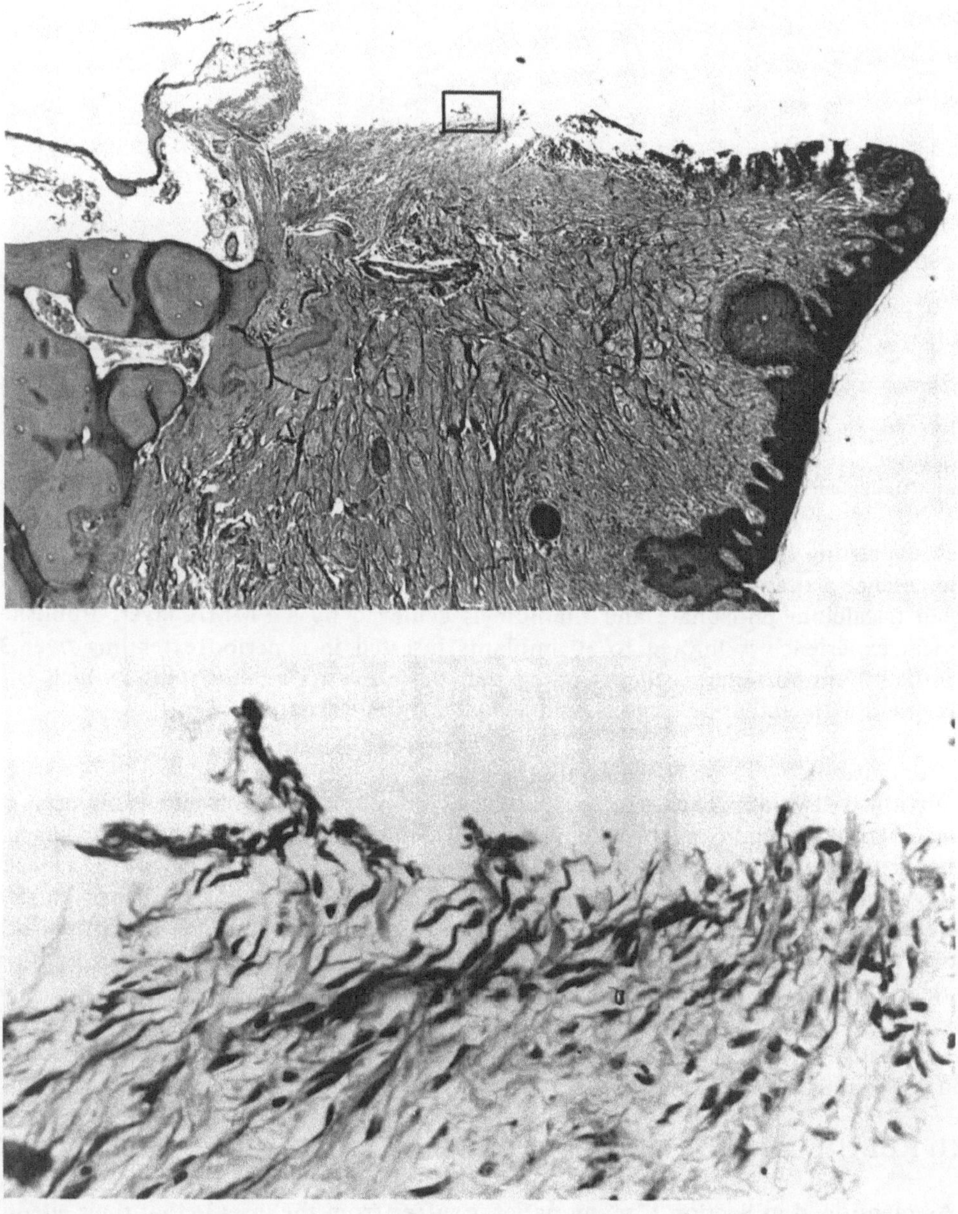

FIGURE 9. Histological evaluation of implants shown in Figure 8. (Top) Left to right shows new bone, connective tissue, epithelial tissue; (bottom) fibers in connective tissue run towards ceramic surface (enlargement of boxed area at bottom).

leading to composite structures, or prestressing, a technique common in the building industry for concrete (Figure 11).

C. Human Applications

Strictly speaking, no human studies have been published where calcium phosphate ceramics have been permucosally evaluated. Riess and coauthors[27] have made use of tricalcium phosphate with an implantable teeth concept, but essentially they covered the intraosseous part of an implant, the "root portion", with this ceramic in order to

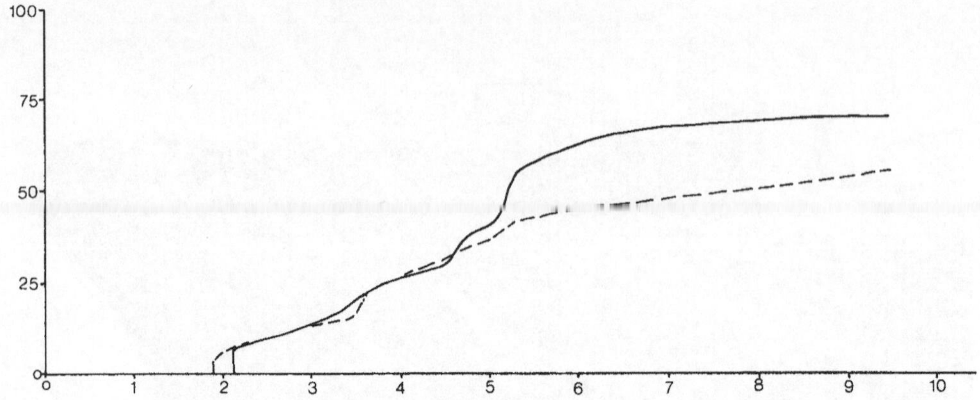

FIGURE 10. (Vertical axis) Percentage of failures due to fracture; (Horizontal axis) Time in months after fixation of implants. Most implants fracture between two and six months after fixation.

ensure a strong ankylotic bond with surrounding alveolar bone. Their "root portion" is in essence a titanium core covered with tricalcium phosphate particles. Bonding between tricalcium phosphate and titanium is achieved by a PMMA layer. Published clinical experience is limited to 30 implants inserted in a period extending over 30 months. Of importance is their finding that the implants remain stable, which is in agreement with studies on dogs as reported later by de Putter et al.[19,20]

D. Conclusions

Obviously, two approaches to the use of calcium phosphate ceramics as material for transmucosal implants have been envisaged. Calcium phosphate can be used to ensure either a tight ankylotic bond alone or both a strong ankylotic bond and a superior gingival attachment. The assumption that calcium phosphate gives a better gingival attachment than other biomaterials is based on several reports in the literature, indicating for example, that hemidesmosomes, specific for gingival attachment to natural teeth, is only found in contact with calcium salts occurring in natural teeth. Nevertheless, a clear-cut study is needed to determine whether calcium phosphate should be used because of its ankylotic bonding alone or of its superior reaction with both bone and gingival tissues.

III. REBUILDING OF DEFICIENT EDENTULOUS ALVEOLAR RIDGES

As mentioned in Section I, many patients suffer from the disease that their edentulous alveolar ridges have been reduced to a point where a satisfying denture cannot be designed anymore. From a physiological point of view, one cannot speak of a pathological process, but of a normal remodeling of the alveolar process. Thus, rebuilding alveolar ridges is not a surgical technique aimed at replacing bone, but at reconstructing bone. As an important conclusion, it follows that resorbable biomaterials are not appropriate, since they will be remodeled again to the "normal", i.e., reduced, state. Experimentally, this conclusion finds its confirmation in the temporary success of autologous bone implants rebuilding reduced ridges.[28]

Clinically, two techniques have been published in the literature, both using nondegradable calcium phosphate ceramics. Kent et al.[29] placed apatite granules alone via a subperiostal tunnel technique, or a mix of apatite with autogenous cancellous bone through a ridge crest incision or lip incision with tunneling. Delivery was done with either technique by making a slurry with blood or saline water. A reported study cov-

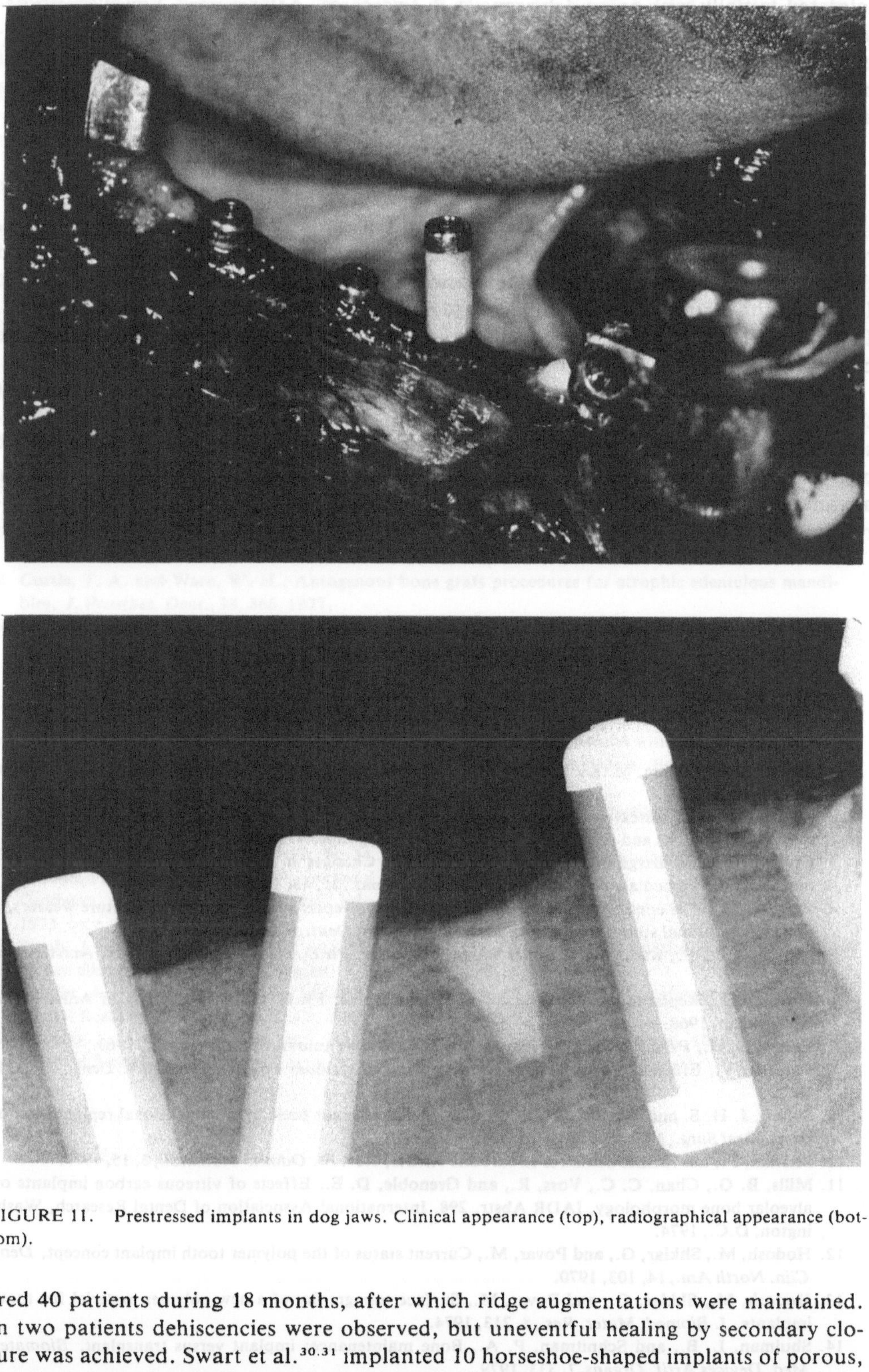

FIGURE 11. Prestressed implants in dog jaws. Clinical appearance (top), radiographical appearance (bottom).

ered 40 patients during 18 months, after which ridge augmentations were maintained. In two patients dehiscencies were observed, but uneventful healing by secondary closure was achieved. Swart et al.[30,31] implanted 10 horseshoe-shaped implants of porous, nondegradable apatite by means of a sandwich technique. These patients observed over four years, showed permanent ridge augmentations as well, and full dentures were

I. INTRODUCTION

Every surgeon dealing with hard tissues is confronted with the problem of repairing bony defects or reconstructing existing bony structures. To perform such work, the surgeon needs extra material which is obtained by removing it from another area of the patient's body. This requires a second operation, and thus doubles the chance of complications. Hence, if there exists a material that can be taken out of a drawer rather than to remove it from the patient's own body, the patient, and the surgeon as well, would be better off.

Other reasons exist, however, to prefer artificially made biomaterials for hard tissue repair or reconstruction:

1. The patient's own bone always participates in remodeling processes, which might be a disadvantage if reconstruction is the goal. It is of no use, for example, to reconstruct a congenital deformation with a remodeling material because the remodeling processes tend to restore the originally present deformation. Remodeling implants should be preferred only for filling defects due to disease or trauma.
2. Sometimes the surgeon needs more material than the patient's own body can provide.
3. The patient's own bone cannot replace all functions in the body, it is not possible to construct a hip joint with bone taken elsewhere from the patient.

Before discussing clinical applications of calcium phosphate ceramics, a review of the use of the patient's own bone ("autologous bone") shall be provided and thereafter follows a review of the limiting properties of calcium phosphate ceramics. Then, the authors will present clinical applications.

II. AUTOLOGOUS BONE GRAFTS

Various articles have been published on bone grafts, an example being the review of Chase and Hendon.[1] Their conclusion, based on reviewing more than 1500 titles, was that present knowledge of the histological fate of bone following transplantation has changed little from views dating from the beginning of the 20th century. Sometimes bone can be transplanted within one wound, but such "local" transplants are only of use for small repairs. Most often, a second wound is necessary, in which case ribs, crista iliaca anterior, crista iliaca posterior, and sometimes tibia bone are preferred. A fresh autologous bone transplant in contact with well-vascularized (bony) tissue of the host site will always be gradually remodeled by a process sometimes called "creeping substitution". The more spongious (cancellous) the transplant, the faster the remodeling.

Autologous bone grafts are usually not very strong from a mechanical point of view. This property renders them useful mainly for sites where no extreme loading occurs. When used for reconstruction, success may not always be expected, as the fate of autologous bone grafts in alveolar ridge augmentation shows (see preceding chapter). Their low strength is illustrated by the fact that by inserting transplants between spinal vertebrae, in order to obtain fusion, crumbling of the implants sometimes occurs, possibly due to resorbtion.[2] The characteristic properties of autologous bone transplants are their limited mechanical strength and the wide variations in remodeling times.[3] Spongious bone is very weak. With compression it collapses at a stress point of 35 kg/cm^2, decreasing with advancing age to less than 20 kg/cm^2 at ages of 60 years and up. Compact bone is much stronger. Transplants obtained from the tibia may have a compressive strength of about 1600 kg/cm^2. Tensile strength values are too low for spongious bone to be measured, and strongly anisotropic for compact bone. When the direction of the stress parallels the osteon direction, values of 800 to 1250 kg/cm^2 are measured, while perpendicular tensile stresses of only 10% of these values are sufficient to fracture a cortical transplant.

Depending on age, individual, and site, a transplant may be replaced by new bone almost completely after one year (>90% for children) or still half may be present after five years (older adult). It cannot be predicted, however, how long it takes to replace bone grafts, so one must conclude that degradation takes place at an unknown rate, varying from more than 100% per year to less than 10% per year. Sometimes roentgenography shows parts of the original tibia transplant after 20 years. A definite disadvantage to this is that obtaining autologous bone transplants requires a second operation, which occasionally results in complications.[3,4]

III. CALCIUM PHOSPHATE CERAMICS: CLINICAL CONSIDERATIONS

A. General Remarks

Possible use of calcium phosphate ceramics in clinical practice depends on the earlier discussed properties. Calcium phosphate ceramics adhere strongly to bone, connective tissue, and (oral) epithelial tissue, can be made biostable or biodegradable (hence, allowing participation in bony remodeling processes) by controlling porosities and (possibly) β-whitlockite content, and are weak in (tensile) fatigue behavior. Thus, the adherence of calcium phosphate ceramics to relevant living tissues equals that of autologous bone transplants. With respect to biodegradability (or the capacity to participate in remodeling), one can conclude that both ceramic and transplant have the capacity, but both with a rate not yet predictable.

Finally, from a biomechanical point of view, spongious bone transplants are definitely weaker than calcium phosphate ceramics, both in tension and compression, while cortical bone has the same compressive strength and a higher tensile strength, depending on direction of tensile stress with respect to osteons. It then follows that these ceramics might be used instead of spongious bone grafts, and replace or reconstruct bone where spongious or cortical bone grafts cannot be applied, but where failure by tensile stresses is not expected.

B. Handling

In handling, the material is most reminiscent of brick. Application of mallet and chisel easily causes fragmentation, and when the material is firmly grasped with forceps some porous material easily breaks or crumbles. The material is best handled with the traditional instruments of orthopedics, oral surgery, and otology: drills, milling cutters, and grindstones, and should be modeled more or less like bone until it has the proper shape and can be introduced. When the material does not fit or is otherwise unsuitable, it can be repeatedly sterilized in gas or the autoclave. Special protective measures, like moistening with Ringer's solution or with serum, are not usually required. When, after implantation, part of the material becomes exposed to air or dehiscencies are found, it is sufficient to remove only the exposed part and to cover the remainder again with adjacent tissue. Removal of the calcium phosphate implant in its entirety is usually not necessary.

C. Infection

Infection is an important factor in implant acceptance. Pathogenic microorganisms may be present prior to introduction of the implant or they may be introduced during the operation. The presence of these microorganisms can give rise to an acute inflam-

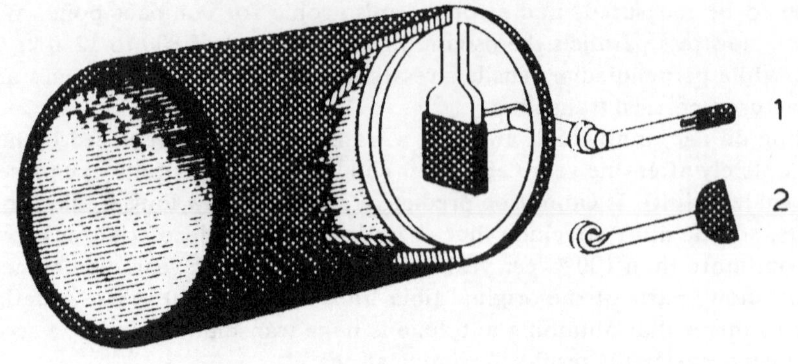

FIGURE 1. Total alloplastic middle ear implant. Solid parts (white) consist of (dental) gold alloy and dashed parts of porous sintered apatite. (1) Apatite piston for contacting stapes footplate; (2) apatite cup-shaped piston for contacting stapes superstructure.

matory process which leads to increasing formation of connective tissue during the healing process, or to chronic inflammatory processes. In these cases, it is a matter of trial and error in deciding on antibiotic medication, and possibly the ultimate therapy might be removal of the implant.

With calcium phosphate implants, infection has hardly been found. However, it cannot be excluded that large, porous calcium phosphate implants may allow infection to build up in the central bulk part, similar to those found by large implants of other materials, such as Proplast®. Pre-impregnating large porous implants with antibiotic fluids may be necessary from a preventive point of view. Another method might be to avoid having pores in the bulk of a large implant. However, in the data presented in this chapter, infection has not yet shown to be a factor of major importance.

IV. CLINICAL USES OF CALCIUM PHOSPHATE CERAMICS

A. Middle Ear Surgery

Sintered hydroxylapatite in reconstructive middle ear surgery has been proposed by Feenstra[5] and Grote.[6] Middle ear surgery of aural cholesteatoma aims at total and safe eradication of the middle ear cavity, since recurrence of the disease occurs otherwise. Reconstruction of a sound conducting system becomes even more difficult, if one realizes that complete eradication of the cholesteatoma is definitely improved by removal of the posterior canal wall.

Grote[7] developed a so-called total alloplastic middle ear prosthesis (TAM), as shown in Figure 1. Specific aspects are

1. Useful in cases with defects in bony anulus and meatal wall as consequences of a safe and total eradication of the diseased middle ear cavity
2. Sound conducting system (ossicular chain) and its suspension (anulus and ear canal) must consist of a monobloc
3. Approximates the normal anatomical and physiological conditions of the middle ear
4. The suspension system and contacting points between ossicular chain and host must be integrated into the host tissue

It is obvious that such a design can only be useful if there is the possibility for a real anchorage into the host. Therefore, Grote started with Proplast® porous hydroxylapatite, since the extrusion rate of Proplast® after longer operative periods is still in debate.

Table 1
ENT IMPLANTATIONS: RESULTS PER 1/1/1981

Patients	Operations	Sex	Age	Follow up (months)	Comments
33—35	Ozaena	3 M	28—30 (29)	3—12 (6)	No problems
36	Orbital roof reconstruction	1 M	40	10	No problems
37—45	Ossicular chain reconstruction	4M, 5 F	14—49 (30)	11—24 (17)	One lost through drum perforation after 2 months
46—49	Posterior canal wall reconstruction	2 M, 2 F	23—50 (41)	11—18 (14)	No problems
50—53	Heightening facial ridge	3 M, 1 F	15—19 (28)	15—19 (18)	No problems
54—65	Partial mastoid obliteration	6 M, 6 F	8—55 (22)	11—19 (16)	One removed because of granulations after 3 months
66—73	Total mastoid obliteration	4 M, 4 F	16—58 (33)	4—15 (10)	No problems
74—75	Rambo type obliteration	2F	46, 31 (39)	14 (14)	One removed because of infection after 3 months
Total		23 M, 20 F	8—58 (29,5)	4—24 (14,2)	

A postoperative observation period of up to two years for two clinical trials is promising, since a hearing gain of 20 dB was claimed. However, the experimental character of this concept requires more long-term clinical studies before evidence can be given whether this TAM is a real alternative in reconstructive surgery.

Feenstra et al.[5] have applied dense hydroxylapatite implants for the reconstruction of the ossicular chain. Apatite stimulates a secure bonding to osseous tissue such as the malleus or to the neotympanum. Twenty such implants functioned well after two years *in situ*. Feenstra also implanted middle ear ventilating tubes of dense calcium phosphate ceramics. These tubes, usually made from plastics, have been used since the early 1960s, and have a few negative side effects. They include tympanosclerosis (possibly due to allergic reactions), and otitis media (a common finding in the allergic patient). Generally, the polyethylene tubes are better than Silastic® PVC tubes as some cause otitis media, and polytetrafluorethylene tubes might give adversary reactions. Therefore, dense apatite ceramics were, not surprisingly, found to have none of these side effects.[8] Table 1 summarizes Feenstra's findings. All these clinical findings on the behavior of apatite ceramics in middle ear surgery agree with those reported on bioglass and glass ceramics having apatite surface layers. Due to the identical surfaces, one would not expect otherwise.

B. Spine Surgery

Ponssen[5] applied macroporous ceramic apatite, resembling spongious autologous 27 implants were inserted in 22 patients having either cervical hernia nuclei pulposi or traumatical instability. Lumbal implants were applied in 10 patients having hernia pulposi. They also were applied in three patients with spondylolisthesis. Table 2 shows the results. Figure 2 gives an example of a successful implant. It is obvious that cervical use of porous apatite implants should be suggested, but that these implants do not

Table 2
SPINE SURGERY: RESULTS OF IMPLANTATIONS PER 1/1/1981

Patients	Intercorporal spondylodesis due to:		Sex	Age	Follow up (months)	Comments
13—14	Traumatic instability	C_{2-3}	M	12	21, 17	No problems
23		C_{3-4}	M	33	9	No problems
15	Traumatic instability	C_{4-5}	F	74	15	No problems
16	Traumatic instability	C_{5-6}				
17	Cervical hernia nucleopulposi	C_{5-6}	M	46	5	No problems
18		C_{6-7}	M	64	7	No problems
19		C_{4-5}				
20		C_{5-6}	M	68	8, 8	No problems
21		C_{5-6}				
22		C_{6-7}	M	54	10,10	Some *initial* radiolucency around implant, ossification after 7 months
23		C_{3-4}	M	42	10	Impression of the implant into vertebral body with secundary fracture after 4 months
24		C_{5-6}	M	62	14	No problems
25		C_{5-6}	M	33	15	No problems
26		C_{4-5}	M	51	16, 16	No problems
27		C_{6-7}	M	42	4	No problems
28		C_{3-4}	F	46	11	No problems
29		C_{6-7}	M	48	10	No problems
30		C_{6-7}	M	47	7	No problems
31		C_{6-7}				
32	Reconstruction defect of frontal skull base		F	8	11	No problems
Total			13 M, 3 F	8—74 (43)	4—21 (9.5)	

have definite advantages as compared with autologous bone for the lumbal region. However, since recent literature[9] suggests that cervical implants made of autologous bone have no advantage compared to not using implants at all, the general conclusion must be that due to a variety of reasons, implants of porous apatite will probably not have a future in spine surgery.

C. Reconstructive Surgery of the Head

According to Swart,[10] developmental deformities of the facial bones can be corrected by osteotomy procedures. This kind of surgery is called orthognatic surgery. Swart described a number of case reports (Table 2) which were followed for about three years, and showed that underdeveloped mandibles can be treated successfully with porous calcium phosphate as well as a hypoplastic maxilla. A chin augmentation procedure, however, was removed after five months, due to persistent infection. Figures 3 and 4 show schematically a Z-shaped osteotomy for lengthening ramus ascendens. In the experience of Swart and Tuinzing,[11] porous apatite ceramics are especially suitable for the Z-shaped osteotomies, as emphasized in a later report.

In all, 15 patients have been successfully treated with porous apatite implants in

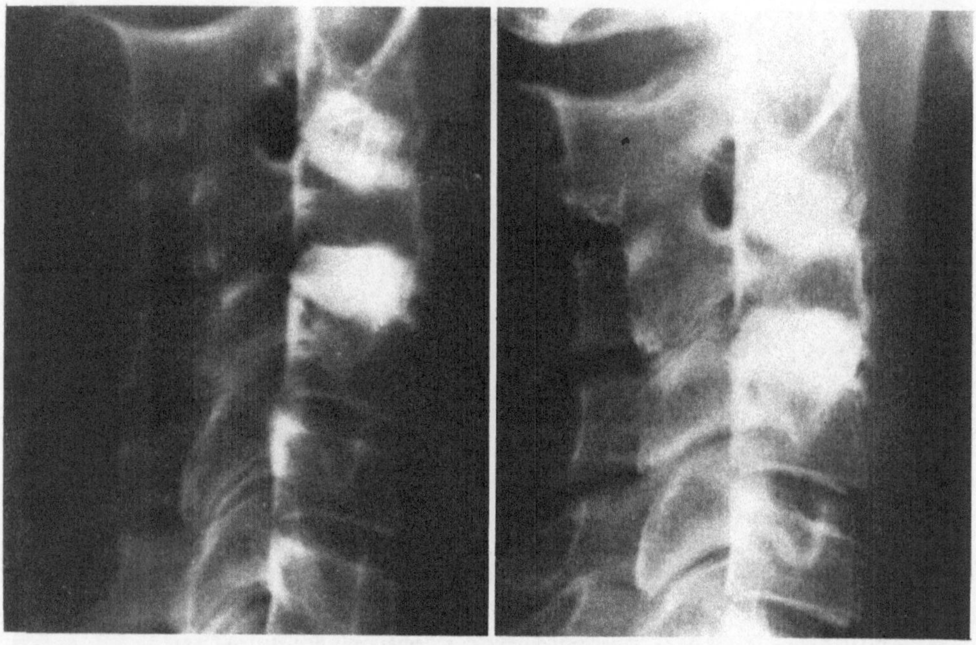

FIGURE 2. Porous apatite implants for fusing vertebrae; (left) X-ray directly after implantation; (right) one year after implantation.

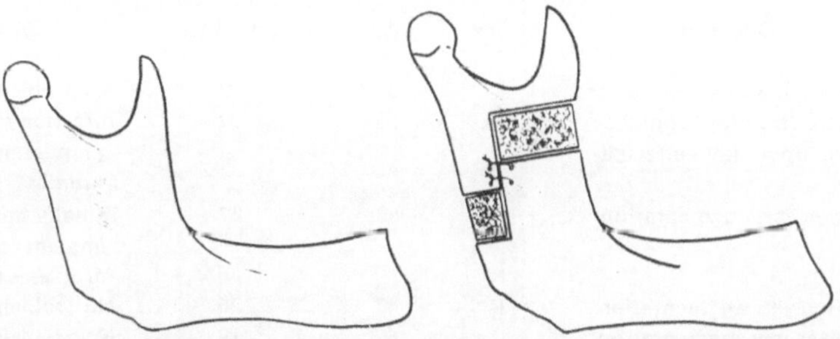

FIGURE 3. Scheme of a Z-shaped osteotomy for lengthening ramus ascendens. Two blocks of porous apatite are needed.

combination with osteotomies. Many (24) implants of this material have been used for various other bone and soft tissue deficiencies in the maxillofacial area (see Table 3). Other than the already mentioned infection of the chin implant (Table 2), one more infection occurred, leading to removal of the implant. Hence, after two to three years, more than 90% (37 of 39) of the implants were still in place causing no discomfort and with a stable result. Walter[12] has treated over 80 patients with maxillofacial problems and found only 5% of the implants to show mild infection, with none leading to a complete loss of the implant. Scraping away the infected area without great change of contours, as visible from the exterior, was sufficient. In contrast with Swart, Walter replaced the air in the porous implants by an antibiotic solution. The data do not allow a conclusion such that an impregnation prevents the need of removing infected implants.

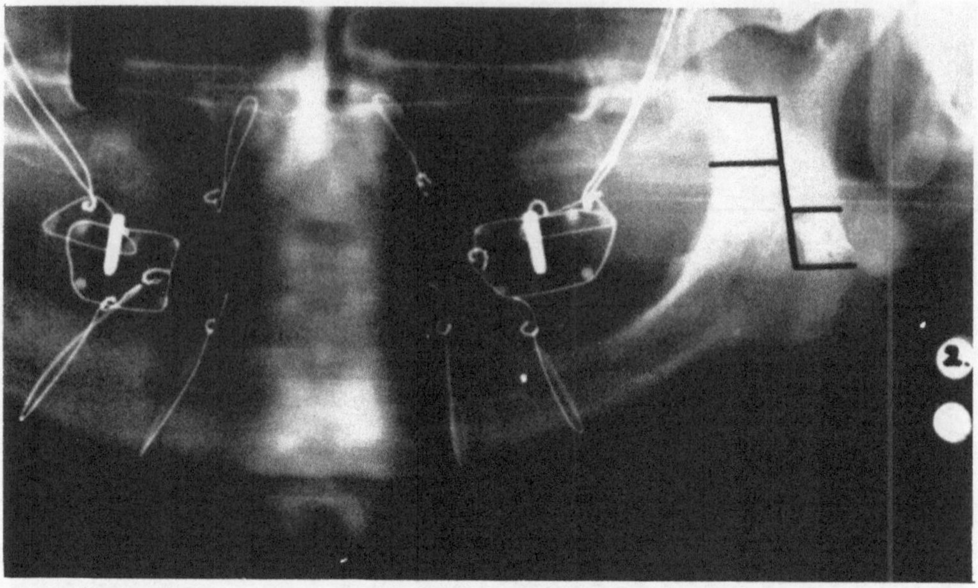

FIGURE 4. X-ray of Z-shaped osteotomy, as schematically depicted in Figure 3.

Table 3
IMPLANTS IN ORAL SURGERY: RESULTS PER 1/1/1981

Patients	Operations	Sex	Age	Follow up (months)	Comments
1	Lower jaw augmentation	F	33	31	No problems
2	Reconstruction frontal defect upper jaw (infected)	M	53	31	Infection remained; implant removed after 1 month
3	Lower jaw augmentation	F	62	27	Initially small piece of implant removed; no problems thereafter
4	Lower jaw augmentation	F	47	26	No problems
5	Lower jaw augmentation	F	67	19	No problems
6	Z-lenghthening of left mandible with implant interposition	M	40	25	Stable contour, unimpaired function
7	Fossa canina augmentation	F	19	18	No problems
8	Chin augmentation	F	21	15	Persistent infection; removal of implant after 5 months
9	Z-lengthening of right mandible with implant interposition	M	30	11	Stable contour; unimpared function
10	Fossa canina augmentation	F	22	7	No problems
11	Fossa canina augmentation	M	23	5	No problems
12	Fossa canina augmentation	F	28	2	No problems
Total		4 M, 8 F	19—67 (37)	2—31 (18)	

D. Filling Defects

Filling defects has been sporadically described. Besides periodontal defects and tooth root implants, as described in the preceding dental chapter, macroporous apatite has

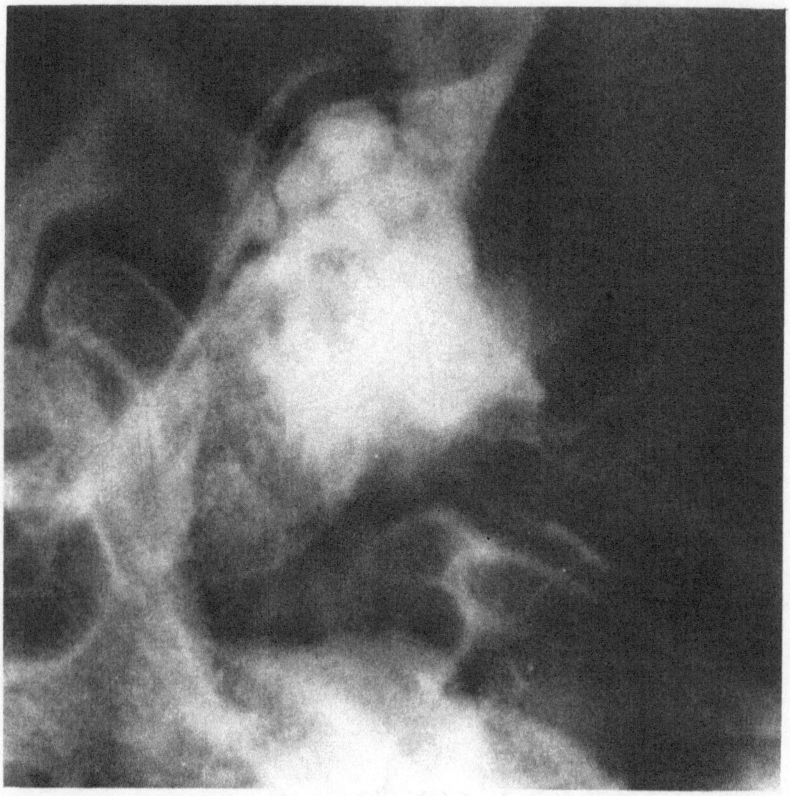

FIGURE 5. X-ray of porous apatite, used to fill space left after a mastoidectomy.

mainly been used to fill mastoidectomies (see Figure 5).[5,13,14] Some patients (23) received such an implant, as described by Feenstra. After an average time of three years, Feenstra reported one complication: the implant was loose without clear signs of infection. Some granulation tissue was formed, however, and a reason might have been that the implant was not fixated well initially, leading to mechanical irritation.

E. Miscellaneous

An interesting application has been suggested by Verschoof[15] who suggested a dumbbell-formed dense apatite implant to obtain a reversible female sterilization (Figure 6). Many efforts have been made to achieve reversible sterilization, since about 5% of the women studied wanted to restore fertility. In addition, there would be a considerable increase in demand for sterilization if reversible procedures were available. One side of the dumbbell occludes the pars intramuralis tubae, while the other side resides in the uterus. This procedure should render laparoscopy unnecessary. Since clinical studies have just started, it remains to be seen whether this device will really be reversible.

The success of permucosal dental implants (see preceding chapter) and ventilating tubes in ENT surgery suggest that calcium phosphate ceramics may succeed as percutaneous implants as well. Surprisingly, no studies in this direction have been reported, but it seems worthwhile to consider calcium phosphate ceramics as a material eliciting acceptable reactions from epidermis when implanted through the skin.

A last "miscellaneous topic" is the possibility of controlled drug release by calcium phosphate implants. The drug can be stored in porous implants, either alone or mixed with an extender (for example, fatty acids).[16] After implantation the release rate of the drug is controlled by the degradation rate of the degradable porous implant or by a diffusion process (out of a nondegradable, porous implant).

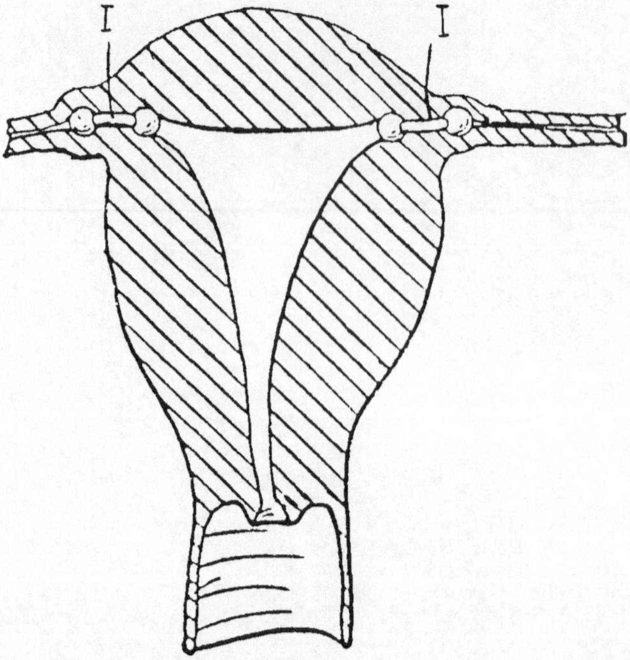

FIGURE 6. Dumbbell formed dense apatite implant (I) to obtain a reversible female sterilization.

V. CONCLUSION

From the preceding discussion it can be concluded that calcium phosphate ceramics have the following applications in medical practice:

1. Their dense form (nondegradable with either apatite or β-whitlockite crystallographic structure) has been successfully used for small, unloaded implants. Replacement of middle ear ossicles, inserting ceramic ventilating tubes, and possible development of a female sterilization device have been discussed.
2. Percutaneous implants made of calcium phosphate ceramics should be considered (for example, for patients with pacemakers or artificial kidney).
3. In reconstructive surgery, especially in the maxillofacial area, nondegradable porous ceramics have shown their efficacy.
4. For filling spaces caused by trauma or disease, porous and degradable ceramics should be the material of choice.
5. Calcium phosphate implants have a possible use as a drug release system: the stored drug may release by degrading the ceramic of by a diffusion process.

In short, calcium phosphate ceramics find use as bone graft substitutes (or extenders), or as reconstructive material if biomechanical forces are absent, with the possible exception of compression.

REFERENCES

1. Chase, S. W. and Hendon, C. H., The fate of autogenous and homogenous bone grafts. An historical review, *J. Bone Jt. Surg.*, 37A, 809, 1955.
2. Hoover, N. W., Methods of lumbar fusion, *J. Bone Jt. Surg.*, 50A, 194, 1968.
3. Gelderman, P. W., The Low Back Pain Syndrome, Thesis, University of Utrecht, Netherlands, 1981.
4. Currey, J. D., The mechanical properties of bone, *Clin. Orthop.*, 59, 125, 1968.
5. Swart, J. G. N., Feenstra, L., Ponssen, H., and de Groot, K., Preliminary clinical experience with sintered hydroxylapatite as a substitute for bone, Ned. Tijdschr. Geneeskd., 123, 1421, 1979.
6. Grote, J. J., Kuypers, W., and de Groot, K., Use of sintered hydroxyapatite in middle ear surgery, *J. Oto. Rhino. Laryngol. Its Borderl.*, 43, 248, 1981.
7. Grote, J. J. and Kuypers, W., Total alloplastic middle ear implant, *Arch. Otolaryngol.*, 106, 560, 1980.
8. Feenstra, L., personal communication.
9. Lunsford, L. B., Bissonette, D. J., Jeannetta, P. J., Sheptak, P. E., and Zorub, D. S., Anterior surgery for cervical disc disease, *J. Neurosurg.*, 53, 1, 1980.
10. Swart, J. G. N. and de Groot, K., Clinical experiences with sintered calciumphosphate as oral implant material, in *Dental Implants*, Heimbe, G., Ed., Carl Hanser Verlag, Munchen, 1980.
11. Tuinzing, D. B. and Swart, T. G. N., Correction of mandibular asymmetry with porous calcium-phosphate implants, *J. Maxillofacial Surg.*, 9, 45, 1981.
12. Walter, C., Internal report, Florence Nightingale Hospital, Dusseldorf-Kaiserswerth, Germany, 1981.
13. Leonard, R. B., Sauer, B. W., and Hulbert, S. F., Use of porous ceramics to obliterate mastoid cavities, *J. Biomed. Mater. Res.*, 4, 85, 1973.
14. Zöllner, C., Strutz, J., Beck, C., Büsing, J. M., Jahnke, K., and Heimke, G., Verödung des Warzen-fortsatzes mit poröser Trikalziumphosphat-Keramik, *Z. Laryng. Rhinol. Otol.*, 60, 85, 1981.
15. Verschoof, K. J. H., Smit, J. W., and de Groot, K., Reversible sterilization: a preliminary report, in press.
16. Reiner, R., Kissing, W., Döring, H., Köster-Lösche, K., and Heide, H., Implantierbares Pharmaka-Depot und Verfahren zu dessen Herstellung, European patent, #0003979, 1981.

INDEX

A

B

C

D

E

F

G

H

I

L

M

N

O

P